MULTIPLIC/

FACTS

A Self-Study Guide

PRACTICE WORKSHEETS

By Shobha

ISBN: 978-0-9997408-1-1

Table of Contents

Before we start...

Knowing multiplication facts is helpful not only in academics; we frequently use multiplication in our daily lives too. Just like learning to walk before you can run, learning multiplication and familiarizing yourself with numbers are building blocks for other math topics taught in school - such as division, long multiplication, fractions and algebra.

Automaticity:

An action that is so well practiced that it does not require conscious effort to carry it out

Mastering the basic math facts develops **automaticity** in kids. Automaticity is the ability to do things without occupying the mind with the low-level details that are required; this is usually the result of consistent learning, repetition, and practice. For instance, an experienced cyclist does not have to concentrate on turning the pedals, balancing, and holding on to the handlebars. Instead, those processes are automatic and the cyclist can concentrate on watching the road, the traffic, and other surroundings.

Until students have developed sufficient sensory-cognitive tools supporting access to symbolic memory, they will not be able to image, store or retrieve all of the basic facts with automaticity. Therefore, students need a comprehensive, developmental, and multi-sensory structured system for developing automaticity with the facts.

The leap from learning subtraction and addition to learning multiplication is one of the most daunting tasks students will face at school, and it's not just students who have trouble with the subject. When teaching multiplication, educators frequently start with the wrong concepts or work through lessons too quickly. This can discourage and intimidate students, ultimately damaging learning outcomes.

Thankfully, there are known strategies to avoid these obstacles.
Our strategy will be a two-step process:

Step 1: Techniques:

Before being tempted to rush and have kids start memorizing facts, we will focus on building a solid foundation by systematically teaching kids what multiplication is by utilizing various techniques and analogies. We will make sure a kid not only knows 2x3 is 6 but also understands why that is a correct answer.

Step 2: Automaticity:

Once the basics are understood we will shift our focus on memorizing the facts to gain automaticity. Memorization will be done with support of several exercises in the worksheets that are targeted to improve speed, consistency and accuracy.

Groups

Before we get into multiplication, it is very important to make sure your child starts seeing numbers in a group as opposed to just "a bunch".

Let's try a simple scenario to understand groups:

Every time Bob's grandmother visits, she brings a box of six donuts.

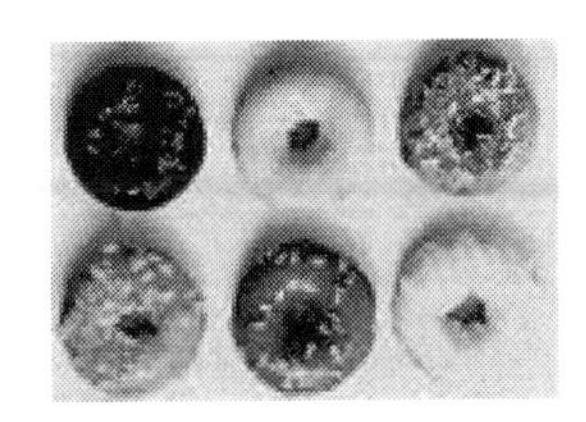

This year, she visited him 3 times.

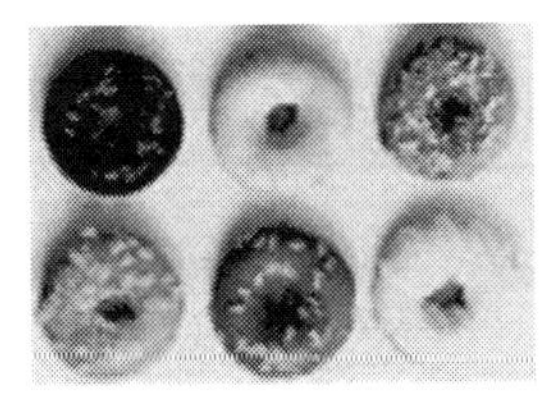

1st visit

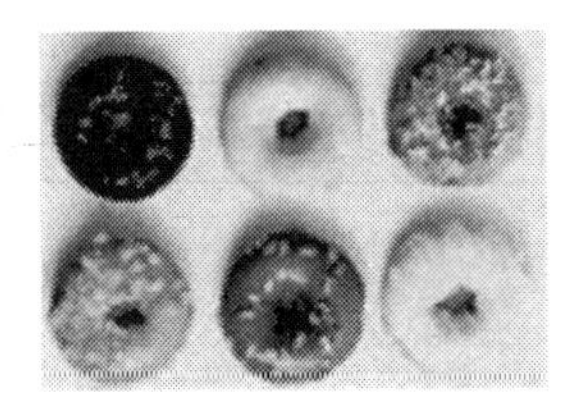

2nd visit

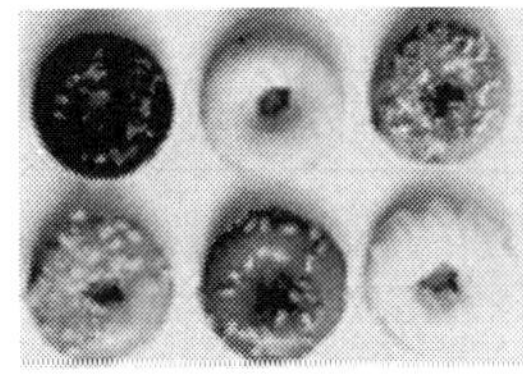

3rd visit

If you count, Bob's grandmother brought a total of 18 donuts in her 3 visits.

Each time she visited, she brought an "**equal sized group**" of 6 donuts

Observe that there are 3 such "equal sized groups".

Instead of counting, we can also use multiplication to find out how many total donuts Bob's grandmother brought.

The symbol for multiplication is "**X**". If we translate this symbol into words it means "**equal sized groups of.**"

X → "equal sized groups of."

There are 3 equal sized groups of 6 donuts.

3 **equal sized groups of** 6 = **3 X 6 = 18**

Let's try some exercises.

Activity 1

Date: .... Start: .... Finish: .... Score: ....

1

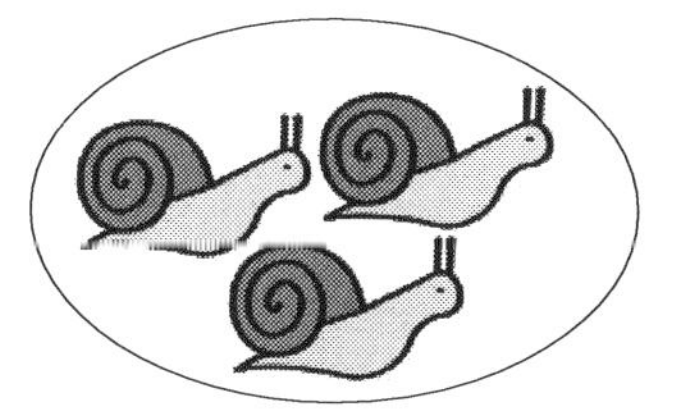

There are ☐ equal sized groups of snails.

Each group has ☐ snails.

Total number of snails = ☐ (groups) x ☐ (in each group) = ☐

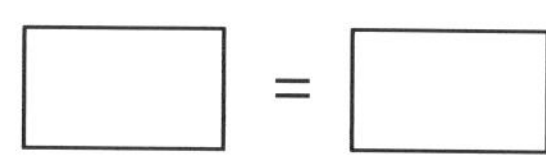

2

There are ☐ equal sized groups of cherries.

Each group has ☐ cherries.

Total number of cherries = ☐ (groups) x ☐ (in each group) = ☐

3

Here each group is a flower with petals.

There are ☐ identical flowers.

Each flower has ☐ petals.

Total number of petals = ☐ (groups) x ☐ (in each group) = ☐

Activity 1

Date: ________ Start: ________ Finish: ________ Score: ________

1

Total number of bunnies = ☐ (groups) x ☐ (in each group) = ☐

2

Total number of trees = ☐ (groups) x ☐ (in each group) = ☐

3

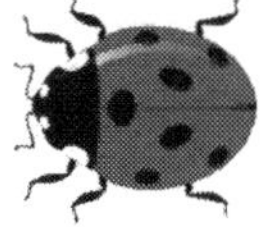  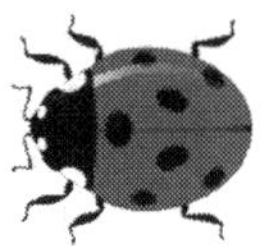

Total number of spots on ladybugs = ☐ (groups) x ☐ (in each group) = ☐

4

Total number of bananas = ☐ (groups) x ☐ (in each group) = ☐

5

Total number of grapes = ☐ (groups) x ☐ (in each group) = ☐

Arrays

An array is an ordered series of objects. Objects in an array are placed in equal sized rows.

Arrays can be used to show multiplication concepts. It helps develop an understanding of the multiplication concepts by visually representing the computation process. The total number of objects in an array can be obtained by counting the total number of rows and number of objects in each row.

Let's try a simple scenario to understand arrays.

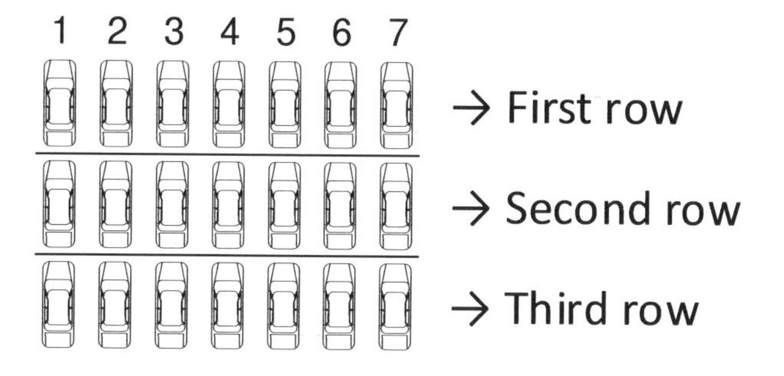

A parking lot can be visually represented as an array with 3 rows with 7 cars in each row. This can be shown by the expression 3 X 7.

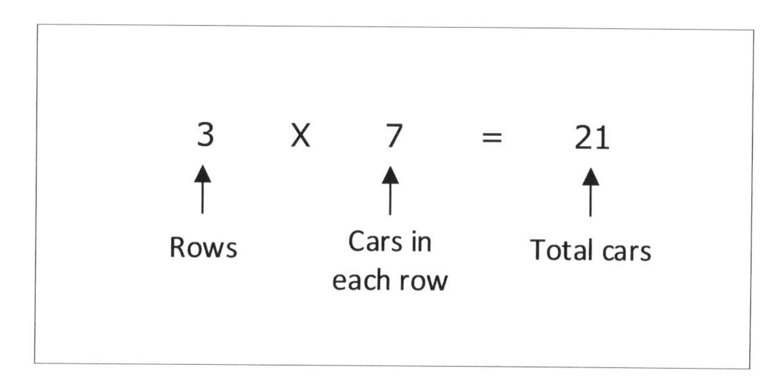

Another example is the egg box shown below. It has 2 rows with 5 eggs in each row.

2 X 5 = 10 total eggs.

Skip Counting

Skip counting (also called number patterns) is counting by a number that is not 1. Using skip counting, we can find the answer to a multiplication problem very easily.

Skip counting is simply counting while skipping a number or numbers in between. In order to skip count, however, the same count needs to be made every time. If counting by 2's you must count every second number each time, for example, 2,4,6,8,10.

This can be represented on the number line.

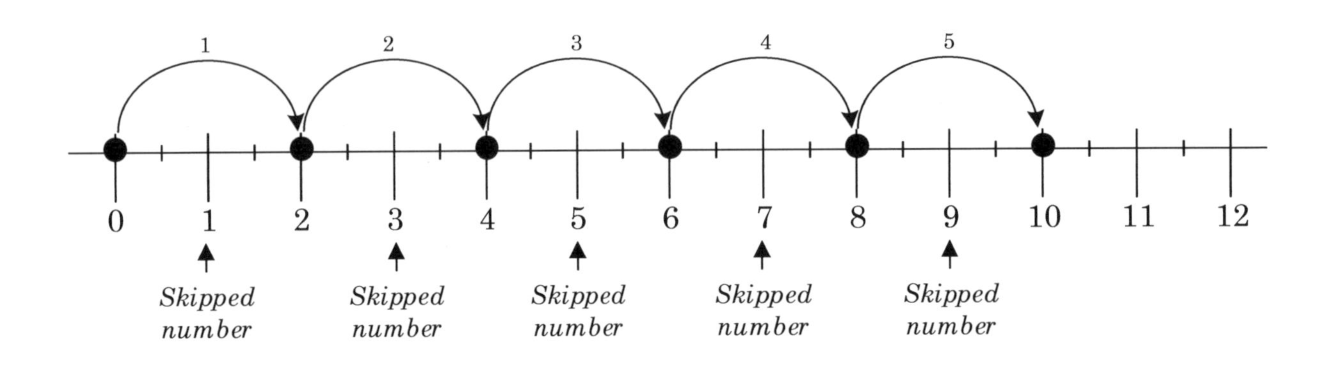

In the above example, 5 jumps of 2 is 10. i.e. 5 X 2 = 10.

Now, look at the example of egg box.

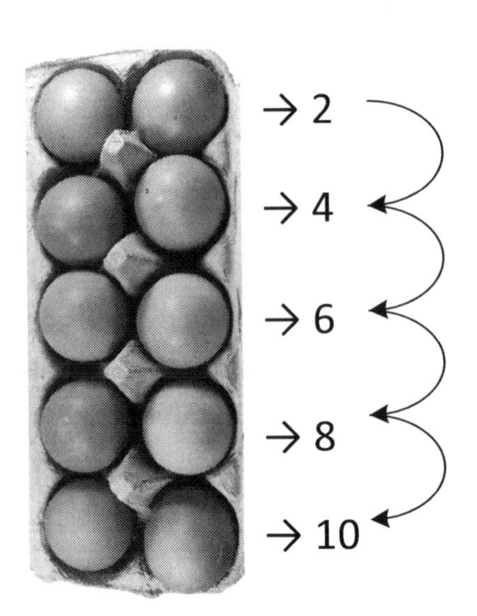

→ 2
→ 4
→ 6
→ 8
→ 10

To find the total number of eggs in the box we can skip count by 2 because each row has 2 eggs.

2 X 5 = 10

Activity 1

Date: ______ Start: ______ Finish: ______ Score: ______

1. Skip count by 2

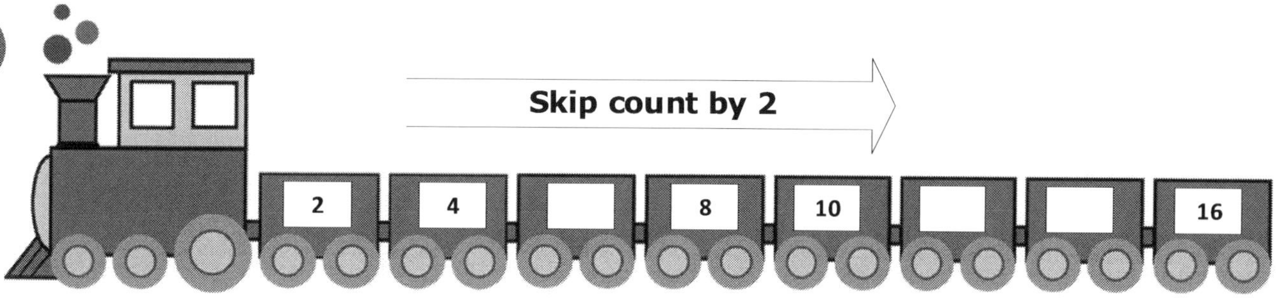

2	4		8	10			16

2. Skip count by 3

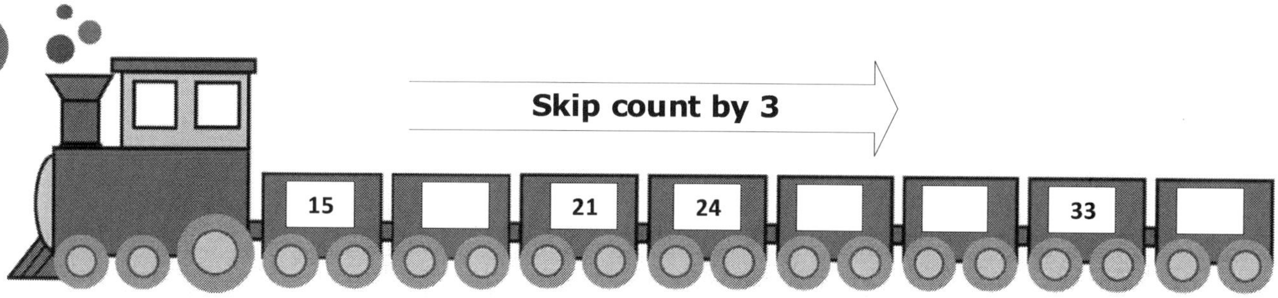

15		21	24			33	

3. Skip count by 4

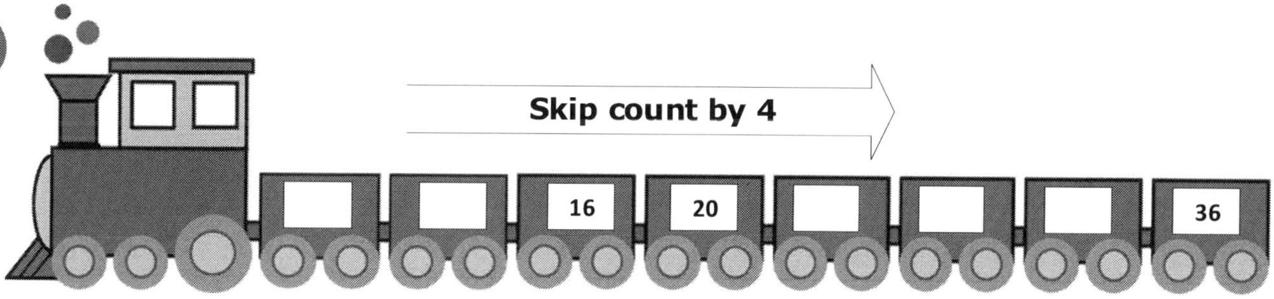

		16	20				36

4. Skip count by 5

	10						40

5. Skip count by 6

18				40			

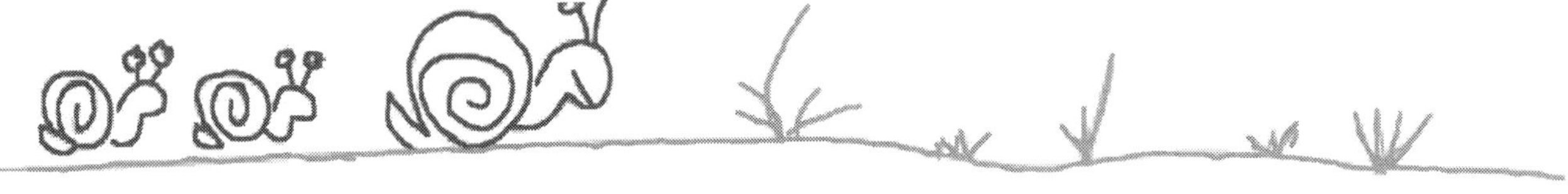

Repeated Addition

Yes, addition and multiplication are related. The concepts you have already learnt about addition is going to help you understand multiplication more lucidly.

Let's go back to the example of Bob's grandmother visits that we used earlier.

We learnt that the symbol "X" indicates multiplication, and multiplication means that you have a certain number of groups of the same size.

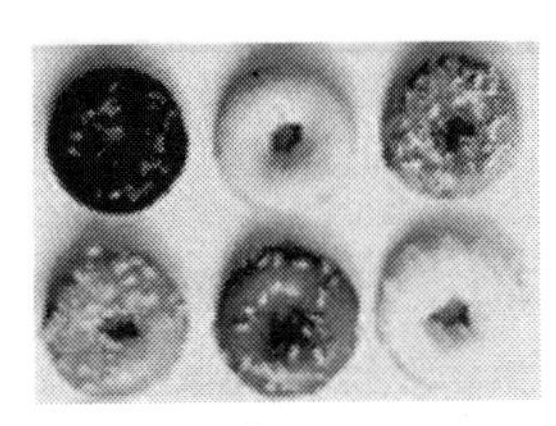

1st visit

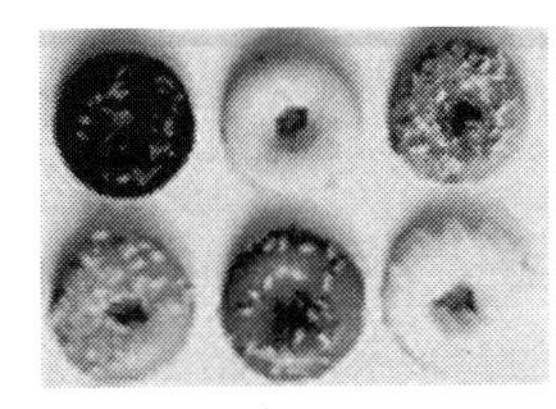

2nd visit

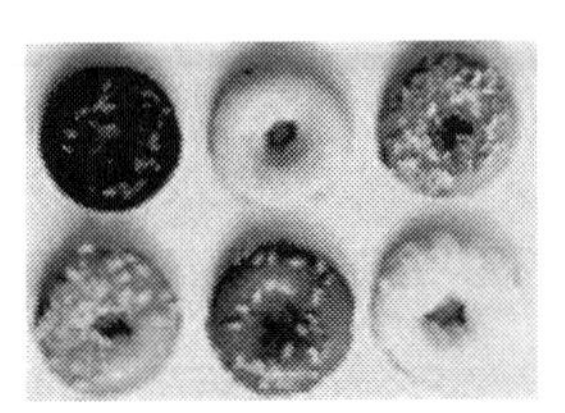

3rd visit

In the above example, there are 3 equal sized groups of 6 donuts. Each of the below expression yields the same result:

3 **equal sized groups of** 6
3 **X** 6
3 **times** 6
6 + 6 + 6

As you can see we can repeatedly add these equal groups to get the total.

Similarly, 5 **X** 8 = 5 **times** 8 = 8 + 8 + 8 + 8 + 8

Activity 1 Date: ________ Start: ________ Finish: ________ Score: ________

1 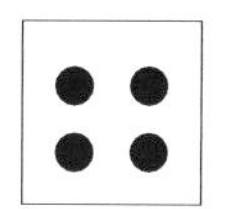 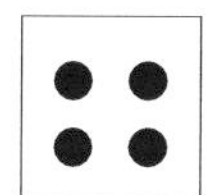

Total number of dots = ___ x ___ = ___ + ___ + ___

2 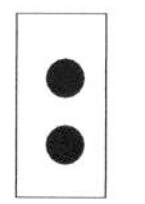    

Total number of dots = ___ x ___ = ___ + ___ + ___ + ___ + ___

3 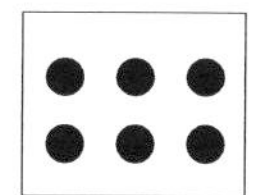 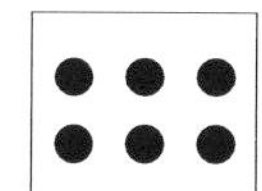

Total number of dots = ___ x ___ = ___ + ___

4 How many legs do four cows have?

Total number of legs = ___ x ___ = ___ + ___ + ___ + ___

5 How many fingers do 3 kids have?

Total number of legs = ___ x ___ = ___ + ___ + ___

6 One row of a marching band has 8 members. How many members are in the band if there are 5 such rows?

Total number of legs = ___ x ___ = ___ + ___ + ___ + ___ + ___

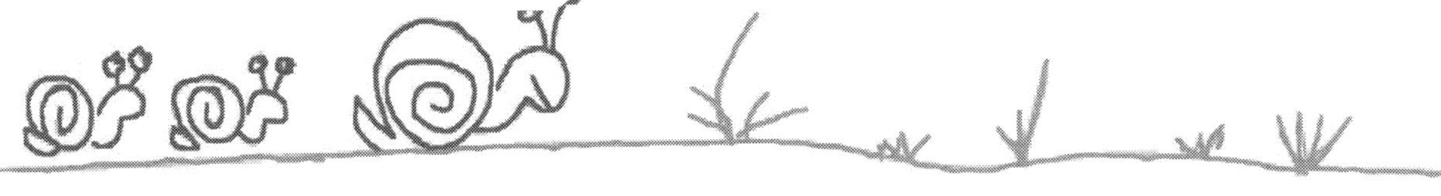

Properties of multiplication

Commutative Property

If we multiply two numbers, it does not matter which number is first or second. They can be multiplied in any order and the result is always the same.

For example, multiplying 4 x 3 will give you the same answer as multiplying 3 x 4.

Let's look at this visually:

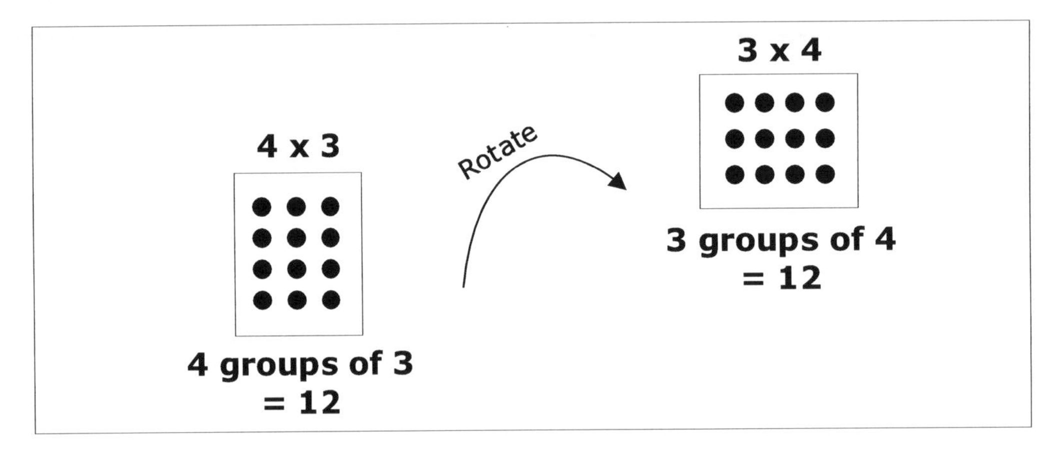

Associative property

It does not matter how we group factors, the product of these factors will always be the same.

For example,

(a x b) x c = a x (b x c)

Or

(3 x 2) x 2 = 3 x (2 x 2) = 12

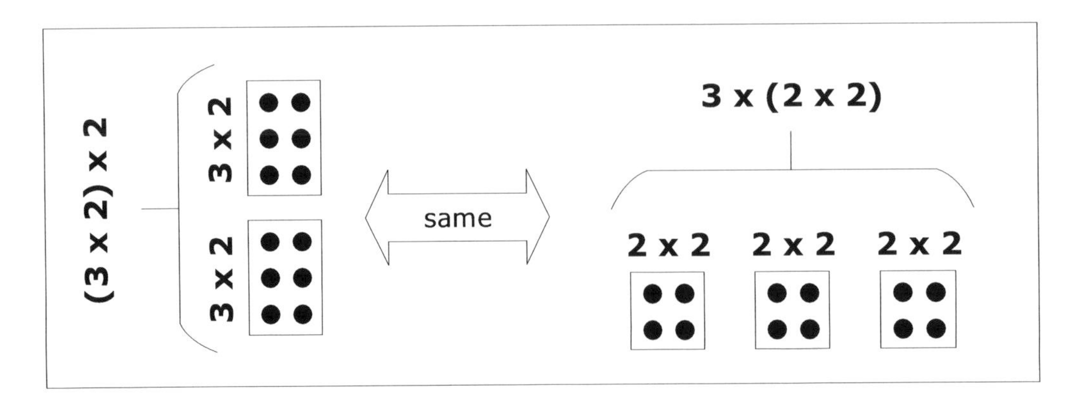

Distributive property

The distributive property lets you multiply a sum by multiplying each addend separately and then add the products.

$$a \times (b + c) = (a \times b) + (a \times c)$$

Or

$$4 \times (3 + 2) = (4 \times 3) + (4 \times 2) = 20$$

In other words, the property spreads out — or, as its name implies, distributes — the value of a equally to b and c.

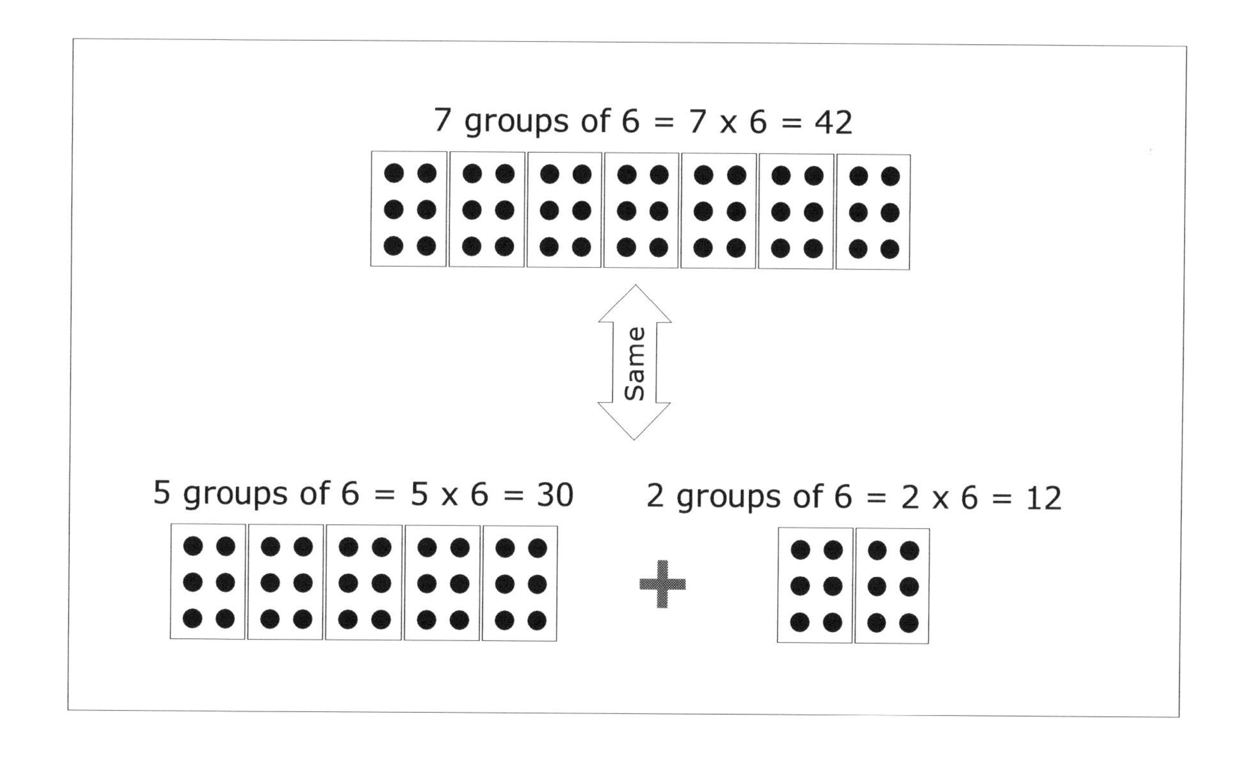

Multiplication Tips and Tricks

Multiplication by 0

If you multiply any number by 0, the result is always 0. For example, 0 x 4 means zero equal sized groups of fours which is 0.

Multiplication by 1

Any number multiplied by 1 always stays the same. For example, 1 X 4 means one group of fours which is nothing but 4.

Multiplication by 2

2 times any number is just double that number. In other words, add the number to itself. For example, 2 x 7 = 7 + 7 = 14.

Multiplication by 3

To multiply by 3, think of a group of 3 as a group of 2 plus 1 more. Since it may be difficult for a child to add 3 numbers simultaneously, first add 2 and then add the third.

Alternatively, you can use your fingers to multiply by 3. Did you notice that each of your finger has three sections? Therefore, you can count out any groups of 3 by counting the sections on each finger. Hold up the number of fingers you're going to multiply by 3. For example, if the problem is 4x3 - hold up four fingers. Count each section on each finger you're holding up, and you should come up with 12.

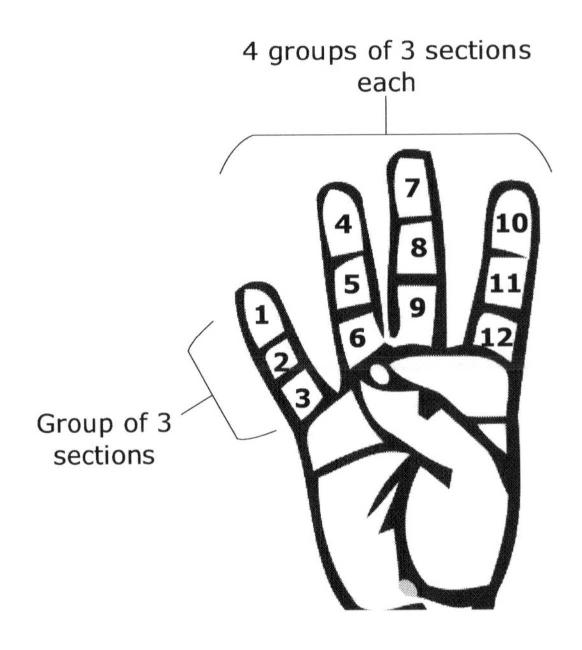

Multiplication Tips and Tricks

Multiplication by 4

To multiply by 4, think of a group of 4 as two groups of 2. So you need to double, then double again.

For example, 4 x 9 is 2 x (2 x 9). First double 9 to get 18 and then double 18 to get 36.

Multiplication by 5

To multiply a number by 5, first multiply by 10 and then cut in half. Since two groups of 5 is 10, a group of 5 is half of a group of 10.

For example, 5 x 8 = 8 times 10 which is 80 then cut in half to get 40.

Multiplication by 6

To multiply a number by 6, think of a group of 6 as two groups of 3. 3 of a number plus 3 of a number is 6 of that number.

6 x 8 or a group of 6 eights is a group of 3 eights plus a group of 3 eights or 24+24 = 48. To multiply by 6, first practice multiplying by 3.

Multiplication by 7

To multiply by 7, think of a group of 7 as a group of 5 plus a group of 2. 5 of a number plus 2 of a number is 7 of that number.

Multiplication by 8

To multiply by 8, double, double and double. Group of 8 is two groups of 4 and each group of 4 is two groups of 2.

For example, 8×4: double 4 is 8, double 8 is 16, double 16 is 32.

Multiplication by 9

To multiply by 9, first multiply by 10 and then subtract the extra one since a group of 9 is one less than a group of 10.

For example, 9×6 = 10×6−6 = 60−6 = 54

Multiplication Table Reference

1 X				
1	×	1	=	1
1	×	2	=	2
1	×	3	=	3
1	×	4	=	4
1	×	5	=	5
1	×	6	=	6
1	×	7	=	7
1	×	8	=	8
1	×	9	=	9
1	×	10	=	10

2 X				
2	×	1	=	2
2	×	2	=	4
2	×	3	=	6
2	×	4	=	8
2	×	5	=	10
2	×	6	=	12
2	×	7	=	14
2	×	8	=	16
2	×	9	=	18
2	×	10	=	20

3 X				
3	×	1	=	3
3	×	2	=	6
3	×	3	=	9
3	×	4	=	12
3	×	5	=	15
3	×	6	=	18
3	×	7	=	21
3	×	8	=	24
3	×	9	=	27
3	×	10	=	30

4 X				
4	×	1	=	4
4	×	2	=	8
4	×	3	=	12
4	×	4	=	16
4	×	5	=	20
4	×	6	=	24
4	×	7	=	28
4	×	8	=	32
4	×	9	=	36
4	×	10	=	40

5 X				
5	×	1	=	5
5	×	2	=	10
5	×	3	=	15
5	×	4	=	20
5	×	5	=	25
5	×	6	=	30
5	×	7	=	35
5	×	8	=	40
5	×	9	=	45
5	×	10	=	50

6 X				
6	×	1	=	6
6	×	2	=	12
6	×	3	=	18
6	×	4	=	24
6	×	5	=	30
6	×	6	=	36
6	×	7	=	42
6	×	8	=	48
6	×	9	=	54
6	×	10	=	60

7 X				
7	×	1	=	7
7	×	2	=	14
7	×	3	=	21
7	×	4	=	28
7	×	5	=	35
7	×	6	=	42
7	×	7	=	49
7	×	8	=	56
7	×	9	=	63
7	×	10	=	70

8 X				
8	×	1	=	8
8	×	2	=	16
8	×	3	=	24
8	×	4	=	32
8	×	5	=	40
8	×	6	=	48
8	×	7	=	56
8	×	8	=	64
8	×	9	=	72
8	×	10	=	80

9 X				
9	×	1	=	9
9	×	2	=	18
9	×	3	=	27
9	×	4	=	36
9	×	5	=	45
9	×	6	=	54
9	×	7	=	63
9	×	8	=	72
9	×	9	=	81
9	×	10	=	90

10 X				
10	×	1	=	10
10	×	2	=	20
10	×	3	=	30
10	×	4	=	40
10	×	5	=	50
10	×	6	=	60
10	×	7	=	70
10	×	8	=	80
10	×	9	=	90
10	×	10	=	100

11 X				
11	×	1	=	11
11	×	2	=	22
11	×	3	=	33
11	×	4	=	44
11	×	5	=	55
11	×	6	=	66
11	×	7	=	77
11	×	8	=	88
11	×	9	=	99
11	×	10	=	110

12 X				
12	×	1	=	12
12	×	2	=	24
12	×	3	=	36
12	×	4	=	48
12	×	5	=	60
12	×	6	=	72
12	×	7	=	84
12	×	8	=	96
12	×	9	=	108
12	×	10	=	120

13 X				
13	×	1	=	13
13	×	2	=	26
13	×	3	=	39
13	×	4	=	52
13	×	5	=	65
13	×	6	=	78
13	×	7	=	91
13	×	8	=	104
13	×	9	=	117
13	×	10	=	130

14 X				
14	×	1	=	14
14	×	2	=	28
14	×	3	=	42
14	×	4	=	56
14	×	5	=	70
14	×	6	=	84
14	×	7	=	98
14	×	8	=	112
14	×	9	=	126
14	×	10	=	140

15 X				
15	×	1	=	15
15	×	2	=	30
15	×	3	=	45
15	×	4	=	60
15	×	5	=	75
15	×	6	=	90
15	×	7	=	105
15	×	8	=	120
15	×	9	=	135
15	×	10	=	150

16 X				
16	×	1	=	16
16	×	2	=	32
16	×	3	=	48
16	×	4	=	64
16	×	5	=	80
16	×	6	=	96
16	×	7	=	112
16	×	8	=	128
16	×	9	=	144
16	×	10	=	160

17 X				
17	×	1	=	17
17	×	2	=	34
17	×	3	=	51
17	×	4	=	68
17	×	5	=	85
17	×	6	=	102
17	×	7	=	119
17	×	8	=	136
17	×	9	=	153
17	×	10	=	170

18 X				
18	×	1	=	18
18	×	2	=	36
18	×	3	=	54
18	×	4	=	72
18	×	5	=	90
18	×	6	=	108
18	×	7	=	126
18	×	8	=	144
18	×	9	=	162
18	×	10	=	180

19 X				
19	×	1	=	19
19	×	2	=	38
19	×	3	=	57
19	×	4	=	76
19	×	5	=	95
19	×	6	=	114
19	×	7	=	133
19	×	8	=	152
19	×	9	=	171
19	×	10	=	190

20 X				
20	×	1	=	20
20	×	2	=	40
20	×	3	=	60
20	×	4	=	80
20	×	5	=	100
20	×	6	=	120
20	×	7	=	140
20	×	8	=	160
20	×	9	=	180
20	×	10	=	200

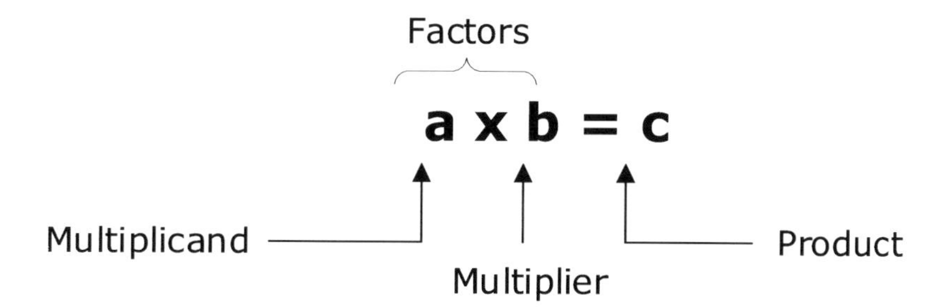

Numbers to be multiplied are called the **multiplier** and the **multiplicand**, or they are sometimes both called "factors." The result of multiplication is called a "product."

Activity 1

Date: ______ Start: ______ Finish: ______ Score: ______

1. 0 X 3 =	2. 1 X 10 =	3. 0 X 5 =
4. 1 X 2 =	5. 0 X 8 =	6. 1 X 6 =
7. 0 X 7 =	8. 1 X 4 =	9. 0 X 9 =
10. 1 X 8 =	11. 0 X 10 =	12. 1 X 2 =
13. 0 X 7 =	14. 1 X 3 =	15. 0 X 6 =
16. 1 X 5 =	17. 0 X 4 =	18. 1 X 9 =
19. 0 X 10 =	20. 1 X 8 =	21. 0 X 6 =

Activity 2

Date: ______ Start: ______ Finish: ______ Score: ______

22. 1 X 7 =	23. 0 X 9 =	24. 1 X 3 =
25. 0 X 2 =	26. 1 X 5 =	27. 0 X 4 =
28. 1 X 2 =	29. 0 X 3 =	30. 1 X 10 =
31. 0 X 9 =	32. 1 X 8 =	33. 0 X 6 =
34. 1 X 7 =	35. 0 X 4 =	36. 1 X 5 =
37. 0 X 7 =	38. 1 X 6 =	39. 0 X 10 =
40. 1 X 8 =	41. 0 X 4 =	42. 1 X 3 =

Activity 1

Date: ________ Start: ________ Finish: ________ Score: ________

1 $2 \times 2 =$ ☐	2 $2 \times 4 =$ ☐	3 $2 \times 3 =$ ☐
4 $2 \times 9 =$ ☐	5 $2 \times 8 =$ ☐	6 $2 \times 6 =$ ☐
7 $2 \times 7 =$ ☐	8 $2 \times 10 =$ ☐	9 $2 \times 5 =$ ☐
10 $2 \times 4 =$ ☐	11 $2 \times 9 =$ ☐	12 $2 \times 8 =$ ☐
13 $2 \times 3 =$ ☐	14 $2 \times 7 =$ ☐	15 $2 \times 2 =$ ☐
16 $2 \times 10 =$ ☐	17 $2 \times 5 =$ ☐	18 $2 \times 6 =$ ☐
19 $2 \times 3 =$ ☐	20 $2 \times 7 =$ ☐	21 $2 \times 8 =$ ☐

Activity 2

Date: ________ Start: ________ Finish: ________ Score: ________

22 $2 \times 10 =$ ☐	23 $2 \times 6 =$ ☐	24 $2 \times 5 =$ ☐
25 $2 \times 2 =$ ☐	26 $2 \times 4 =$ ☐	27 $2 \times 9 =$ ☐
28 $2 \times 5 =$ ☐	29 $2 \times 9 =$ ☐	30 $2 \times 4 =$ ☐
31 $2 \times 3 =$ ☐	32 $2 \times 6 =$ ☐	33 $2 \times 2 =$ ☐
34 $2 \times 10 =$ ☐	35 $2 \times 7 =$ ☐	36 $2 \times 8 =$ ☐
37 $2 \times 5 =$ ☐	38 $2 \times 8 =$ ☐	39 $2 \times 3 =$ ☐
40 $2 \times 2 =$ ☐	41 $2 \times 6 =$ ☐	42 $2 \times 9 =$ ☐

Practice: Multiplication Table 2

Activity 1

Date: ________ Start: ________ Finish: ________ Score: ________

1. 2 X 6 = ☐
2. 2 X 10 = ☐
3. 2 X 3 = ☐
4. 2 X 4 = ☐
5. 2 X 8 = ☐
6. 2 X 9 = ☐
7. 2 X 7 = ☐
8. 2 X 5 = ☐
9. 2 X 2 = ☐
10. 2 X 3 = ☐
11. 2 X 2 = ☐
12. 2 X 4 = ☐
13. 2 X 8 = ☐
14. 2 X 10 = ☐
15. 2 X 9 = ☐
16. 2 X 6 = ☐
17. 2 X 5 = ☐
18. 2 X 7 = ☐
19. 2 X 9 = ☐
20. 2 X 7 = ☐
21. 2 X 8 = ☐

Activity 2

Date: ________ Start: ________ Finish: ________ Score: ________

22. 2 X 4 = ☐
23. 2 X 2 = ☐
24. 2 X 10 = ☐
25. 2 X 6 = ☐
26. 2 X 3 = ☐
27. 2 X 5 = ☐
28. 2 X 2 = ☐
29. 2 X 8 = ☐
30. 2 X 7 = ☐
31. 2 X 9 = ☐
32. 2 X 10 = ☐
33. 2 X 5 = ☐
34. 2 X 6 = ☐
35. 2 X 4 = ☐
36. 2 X 3 = ☐
37. 2 X 9 = ☐
38. 2 X 6 = ☐
39. 2 X 2 = ☐
40. 2 X 4 = ☐
41. 2 X 8 = ☐
42. 2 X 5 = ☐

Activity 1

Date: ______ Start: ______ Finish: ______ Score: ______

1. 2 X 3 = ____	2. 2 X 7 = ____	3. 2 X 10 = ____
4. 2 X 5 = ____	5. 2 X 4 = ____	6. 2 X 8 = ____
7. 2 X 2 = ____	8. 2 X 9 = ____	9. 2 X 6 = ____
10. 2 X 2 = ____	11. 2 X 8 = ____	12. 2 X 9 = ____
13. 2 X 6 = ____	14. 2 X 3 = ____	15. 2 X 10 = ____
16. 2 X 4 = ____	17. 2 X 7 = ____	18. 2 X 5 = ____
19. 2 X 8 = ____	20. 2 X 5 = ____	21. 2 X 9 = ____

Activity 2

Date: ______ Start: ______ Finish: ______ Score: ______

22. 2 X 10 = ____	23. 2 X 2 = ____	24. 2 X 6 = ____
25. 2 X 7 = ____	26. 2 X 4 = ____	27. 2 X 3 = ____
28. 2 X 5 = ____	29. 2 X 4 = ____	30. 2 X 6 = ____
31. 2 X 7 = ____	32. 2 X 10 = ____	33. 2 X 3 = ____
34. 2 X 8 = ____	35. 2 X 2 = ____	36. 2 X 9 = ____
37. 2 X 10 = ____	38. 2 X 2 = ____	39. 2 X 9 = ____
40. 2 X 7 = ____	41. 2 X 8 = ____	42. 2 X 3 = ____

Practice: Multiplication Table 3

Activity 1

Date: ______ Start: ______ Finish: ______ Score: ______

1	3 X 9 = ☐	2	3 X 6 = ☐	3	3 X 3 = ☐
4	3 X 10 = ☐	5	3 X 8 = ☐	6	3 X 7 = ☐
7	3 X 5 = ☐	8	3 X 4 = ☐	9	3 X 2 = ☐
10	3 X 6 = ☐	11	3 X 7 = ☐	12	3 X 4 = ☐
13	3 X 3 = ☐	14	3 X 10 = ☐	15	3 X 2 = ☐
16	3 X 5 = ☐	17	3 X 9 = ☐	18	3 X 8 = ☐
19	3 X 9 = ☐	20	3 X 8 = ☐	21	3 X 10 = ☐

Activity 2

Date: ______ Start: ______ Finish: ______ Score: ______

22	3 X 5 = ☐	23	3 X 6 = ☐	24	3 X 2 = ☐
25	3 X 7 = ☐	26	3 X 4 = ☐	27	3 X 3 = ☐
28	3 X 5 = ☐	29	3 X 10 = ☐	30	3 X 4 = ☐
31	3 X 8 = ☐	32	3 X 9 = ☐	33	3 X 7 = ☐
34	3 X 6 = ☐	35	3 X 3 = ☐	36	3 X 2 = ☐
37	3 X 3 = ☐	38	3 X 8 = ☐	39	3 X 9 = ☐
40	3 X 6 = ☐	41	3 X 4 = ☐	42	3 X 10 = ☐

Activity 1

Date: ______ Start: ______ Finish: ______ Score: ______

1 $3 \times 9 =$ ☐	2 $3 \times 3 =$ ☐	3 $3 \times 5 =$ ☐
4 $3 \times 4 =$ ☐	5 $3 \times 10 =$ ☐	6 $3 \times 2 =$ ☐
7 $3 \times 7 =$ ☐	8 $3 \times 6 =$ ☐	9 $3 \times 8 =$ ☐
10 $3 \times 5 =$ ☐	11 $3 \times 2 =$ ☐	12 $3 \times 10 =$ ☐
13 $3 \times 7 =$ ☐	14 $3 \times 4 =$ ☐	15 $3 \times 9 =$ ☐
16 $3 \times 6 =$ ☐	17 $3 \times 3 =$ ☐	18 $3 \times 8 =$ ☐
19 $3 \times 5 =$ ☐	20 $3 \times 7 =$ ☐	21 $3 \times 4 =$ ☐

Activity 2

Date: ______ Start: ______ Finish: ______ Score: ______

22 $3 \times 3 =$ ☐	23 $3 \times 2 =$ ☐	24 $3 \times 10 =$ ☐
25 $3 \times 8 =$ ☐	26 $3 \times 6 =$ ☐	27 $3 \times 9 =$ ☐
28 $3 \times 10 =$ ☐	29 $3 \times 7 =$ ☐	30 $3 \times 8 =$ ☐
31 $3 \times 2 =$ ☐	32 $3 \times 9 =$ ☐	33 $3 \times 6 =$ ☐
34 $3 \times 4 =$ ☐	35 $3 \times 5 =$ ☐	36 $3 \times 3 =$ ☐
37 $3 \times 6 =$ ☐	38 $3 \times 4 =$ ☐	39 $3 \times 9 =$ ☐
40 $3 \times 3 =$ ☐	41 $3 \times 7 =$ ☐	42 $3 \times 5 =$ ☐

Practice: Multiplication Table 3

Activity 1

Date: ________ Start: ________ Finish: ________ Score: ________

1 3 X 9 = ☐	2 3 X 7 = ☐	3 3 X 5 = ☐
4 3 X 2 = ☐	5 3 X 10 = ☐	6 3 X 8 = ☐
7 3 X 6 = ☐	8 3 X 3 = ☐	9 3 X 4 = ☐
10 3 X 2 = ☐	11 3 X 6 = ☐	12 3 X 5 = ☐
13 3 X 10 = ☐	14 3 X 3 = ☐	15 3 X 8 = ☐
16 3 X 4 = ☐	17 3 X 7 = ☐	18 3 X 9 = ☐
19 3 X 4 = ☐	20 3 X 3 = ☐	21 3 X 5 = ☐

Activity 2

Date: ________ Start: ________ Finish: ________ Score: ________

22 3 X 9 = ☐	23 3 X 7 = ☐	24 3 X 8 = ☐
25 3 X 2 = ☐	26 3 X 10 = ☐	27 3 X 6 = ☐
28 3 X 4 = ☐	29 3 X 8 = ☐	30 3 X 5 = ☐
31 3 X 10 = ☐	32 3 X 7 = ☐	33 3 X 3 = ☐
34 3 X 6 = ☐	35 3 X 2 = ☐	36 3 X 9 = ☐
37 3 X 10 = ☐	38 3 X 7 = ☐	39 3 X 6 = ☐
40 3 X 9 = ☐	41 3 X 8 = ☐	42 3 X 2 = ☐

Activity 1

Date: ______ Start: ______ Finish: ______ Score: ______

1	4 X 5 =	2	4 X 4 =	3	4 X 9 =
4	4 X 10 =	5	4 X 8 =	6	4 X 2 =
7	4 X 6 =	8	4 X 7 =	9	4 X 3 =
10	4 X 10 =	11	4 X 3 =	12	4 X 4 =
13	4 X 7 =	14	4 X 9 =	15	4 X 8 =
16	4 X 6 =	17	4 X 5 =	18	4 X 2 =
19	4 X 5 =	20	4 X 6 =	21	4 X 4 =

Activity 2

Date: ______ Start: ______ Finish: ______ Score: ______

22	4 X 9 =	23	4 X 10 =	24	4 X 3 =
25	4 X 7 =	26	4 X 2 =	27	4 X 8 =
28	4 X 7 =	29	4 X 2 =	30	4 X 4 =
31	4 X 8 =	32	4 X 5 =	33	4 X 9 =
34	4 X 6 =	35	4 X 3 =	36	4 X 10 =
37	4 X 7 =	38	4 X 3 =	39	4 X 5 =
40	4 X 6 =	41	4 X 4 =	42	4 X 9 =

Activity 1

Date: ________ Start: ________ Finish: ________ Score: ________

1. $4 \times 2 =$ ☐	2. $4 \times 9 =$ ☐	3. $4 \times 5 =$ ☐
4. $4 \times 10 =$ ☐	5. $4 \times 4 =$ ☐	6. $4 \times 7 =$ ☐
7. $4 \times 6 =$ ☐	8. $4 \times 3 =$ ☐	9. $4 \times 8 =$ ☐
10. $4 \times 3 =$ ☐	11. $4 \times 2 =$ ☐	12. $4 \times 8 =$ ☐
13. $4 \times 10 =$ ☐	14. $4 \times 4 =$ ☐	15. $4 \times 6 =$ ☐
16. $4 \times 5 =$ ☐	17. $4 \times 7 =$ ☐	18. $4 \times 9 =$ ☐
19. $4 \times 2 =$ ☐	20. $4 \times 7 =$ ☐	21. $4 \times 8 =$ ☐

Activity 2

Date: ________ Start: ________ Finish: ________ Score: ________

22. $4 \times 10 =$ ☐	23. $4 \times 4 =$ ☐	24. $4 \times 9 =$ ☐
25. $4 \times 6 =$ ☐	26. $4 \times 3 =$ ☐	27. $4 \times 5 =$ ☐
 28. $4 \times 8 =$ ☐	29. $4 \times 5 =$ ☐	30. $4 \times 10 =$ ☐
31. $4 \times 6 =$ ☐	32. $4 \times 4 =$ ☐	33. $4 \times 3 =$ ☐
34. $4 \times 9 =$ ☐	35. $4 \times 2 =$ ☐	36. $4 \times 7 =$ ☐
37. $4 \times 3 =$ ☐	38. $4 \times 4 =$ ☐	39. $4 \times 7 =$ ☐
40. $4 \times 6 =$ ☐	41. $4 \times 9 =$ ☐	42. $4 \times 10 =$ ☐

Activity 1

Date: ________ Start: ________ Finish: ________ Score: ________

1. 4 X 2 = ☐
2. 4 X 7 = ☐
3. 4 X 4 = ☐
4. 4 X 9 = ☐
5. 4 X 5 = ☐
6. 4 X 3 = ☐
7. 4 X 8 = ☐
8. 4 X 10 = ☐
9. 4 X 6 = ☐
10. 4 X 7 = ☐
11. 4 X 6 = ☐
12. 4 X 9 = ☐
13. 4 X 3 = ☐
14. 4 X 4 = ☐
15. 4 X 2 = ☐
16. 4 X 10 = ☐
17. 4 X 5 = ☐
18. 4 X 8 = ☐
19. 4 X 7 = ☐
20. 4 X 10 = ☐
21. 4 X 6 = ☐

Activity 2

Date: ________ Start: ________ Finish: ________ Score: ________

22. 4 X 9 = ☐
23. 4 X 3 = ☐
24. 4 X 2 = ☐
25. 4 X 8 = ☐
26. 4 X 5 = ☐
27. 4 X 4 = ☐
28. 4 X 5 = ☐
29. 4 X 9 = ☐
30. 4 X 7 = ☐
31. 4 X 3 = ☐
32. 4 X 2 = ☐
33. 4 X 4 = ☐
34. 4 X 10 = ☐
35. 4 X 6 = ☐
36. 4 X 8 = ☐
37. 4 X 10 = ☐
38. 4 X 7 = ☐
39. 4 X 2 = ☐
40. 4 X 6 = ☐
41. 4 X 4 = ☐
42. 4 X 9 = ☐

Practice: Multiplication Table 5

Activity 1

Date: ______ Start: ______ Finish: ______ Score: ______

1. 5 X 5 = ☐	2. 5 X 8 = ☐	3. 5 X 7 = ☐
4. 5 X 3 = ☐	5. 5 X 4 = ☐	6. 5 X 2 = ☐
7. 5 X 10 = ☐	8. 5 X 9 = ☐	9. 5 X 6 = ☐
10. 5 X 3 = ☐	11. 5 X 7 = ☐	12. 5 X 8 = ☐
13. 5 X 5 = ☐	14. 5 X 9 = ☐	15. 5 X 6 = ☐
16. 5 X 4 = ☐	17. 5 X 10 = ☐	18. 5 X 2 = ☐
19. 5 X 3 = ☐	20. 5 X 10 = ☐	21. 5 X 6 = ☐

Activity 2

Date: ______ Start: ______ Finish: ______ Score: ______

22. 5 X 5 = ☐	23. 5 X 8 = ☐	24. 5 X 4 = ☐
25. 5 X 7 = ☐	26. 5 X 9 = ☐	27. 5 X 2 = ☐
28. 5 X 9 = ☐	29. 5 X 10 = ☐	30. 5 X 7 = ☐
31. 5 X 2 = ☐	32. 5 X 8 = ☐	33. 5 X 3 = ☐
34. 5 X 4 = ☐	35. 5 X 6 = ☐	36. 5 X 5 = ☐
37. 5 X 2 = ☐	38. 5 X 6 = ☐	39. 5 X 3 = ☐
40. 5 X 9 = ☐	41. 5 X 7 = ☐	42. 5 X 10 = ☐

Activity 1

Date: ________ Start: ________ Finish: ________ Score: ________

1. 5 X 3 =	2. 5 X 5 =	3. 5 X 6 =
4. 5 X 8 =	5. 5 X 2 =	6. 5 X 7 =
7. 5 X 9 =	8. 5 X 10 =	9. 5 X 4 =
10. 5 X 7 =	11. 5 X 10 =	12. 5 X 4 =
13. 5 X 3 =	14. 5 X 6 =	15. 5 X 5 =
16. 5 X 8 =	17. 5 X 2 =	18. 5 X 9 =
19. 5 X 5 =	20. 5 X 7 =	21. 5 X 9 =

Activity 2

Date: ________ Start: ________ Finish: ________ Score: ________

22. 5 X 2 =	23. 5 X 3 =	24. 5 X 10 =
25. 5 X 6 =	26. 5 X 8 =	27. 5 X 4 =
28. 5 X 5 =	29. 5 X 2 =	30. 5 X 6 =
31. 5 X 7 =	32. 5 X 4 =	33. 5 X 9 =
34. 5 X 10 =	35. 5 X 3 =	36. 5 X 8 =
37. 5 X 10 =	38. 5 X 9 =	39. 5 X 6 =
40. 5 X 7 =	41. 5 X 2 =	42. 5 X 3 =

Practice: Multiplication Table 5

Activity 1

Date: ________ Start: ________ Finish: ________ Score: ________

1	5 X 5 =		2	5 X 8 =		3	5 X 2 =	
4	5 X 7 =		5	5 X 4 =		6	5 X 9 =	
7	5 X 3 =		8	5 X 10 =		9	5 X 6 =	
10	5 X 9 =		11	5 X 10 =		12	5 X 7 =	
13	5 X 5 =		14	5 X 6 =		15	5 X 2 =	
16	5 X 4 =		17	5 X 3 =		18	5 X 8 =	
19	5 X 3 =		20	5 X 8 =		21	5 X 7 =	

Activity 2

Date: ________ Start: ________ Finish: ________ Score: ________

22	5 X 2 =		23	5 X 9 =		24	5 X 4 =	
25	5 X 5 =		26	5 X 6 =		27	5 X 10 =	
28	5 X 5 =		29	5 X 7 =		30	5 X 2 =	
31	5 X 6 =		32	5 X 8 =		33	5 X 10 =	
34	5 X 9 =		35	5 X 3 =		36	5 X 4 =	
37	5 X 6 =		38	5 X 8 =		39	5 X 4 =	
40	5 X 9 =		41	5 X 5 =		42	5 X 7 =	

Activity 1

Date: ______ Start: ______ Finish: ______ Score: ______

1. 3 X 9 = ☐	2. 4 X 5 = ☐	3. 5 X 10 = ☐
4. 2 X 4 = ☐	5. 5 X 3 = ☐	6. 3 X 2 = ☐
7. 4 X 7 = ☐	8. 2 X 8 = ☐	9. 3 X 6 = ☐
10. 2 X 10 = ☐	11. 5 X 2 = ☐	12. 4 X 6 = ☐
13. 2 X 7 = ☐	14. 4 X 3 = ☐	15. 5 X 4 = ☐
16. 3 X 9 = ☐	17. 4 X 8 = ☐	18. 2 X 5 = ☐
19. 5 X 7 = ☐	20. 3 X 9 = ☐	21. 4 X 3 = ☐

Activity 2

Date: ______ Start: ______ Finish: ______ Score: ______

22. 5 X 6 = ☐	23. 2 X 5 = ☐	24. 3 X 8 = ☐
25. 5 X 10 = ☐	26. 4 X 4 = ☐	27. 3 X 2 = ☐
28. 2 X 8 = ☐	29. 5 X 9 = ☐	30. 3 X 4 = ☐
31. 4 X 6 = ☐	32. 2 X 7 = ☐	33. 3 X 5 = ☐
34. 2 X 10 = ☐	35. 5 X 2 = ☐	36. 4 X 3 = ☐
37. 2 X 6 = ☐	38. 3 X 2 = ☐	39. 4 X 9 = ☐
40. 5 X 10 = ☐	41. 4 X 5 = ☐	42. 2 X 3 = ☐

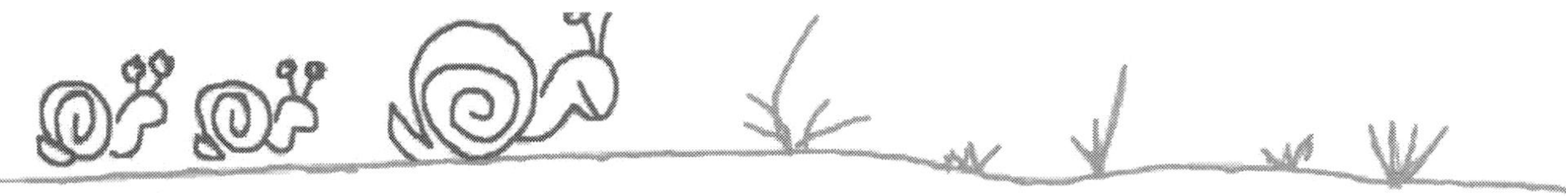

Activity 1

Date: ______ Start: ______ Finish: ______ Score: ______

1. 3 X 5 =	2. 4 X 8 =	3. 5 X 4 =
4. 2 X 3 =	5. 4 X 7 =	6. 2 X 10 =
7. 3 X 9 =	8. 5 X 6 =	9. 3 X 2 =
10. 5 X 3 =	11. 4 X 7 =	12. 2 X 2 =
13. 4 X 6 =	14. 3 X 8 =	15. 5 X 4 =
16. 2 X 10 =	17. 5 X 5 =	18. 2 X 9 =
19. 4 X 10 =	20. 3 X 3 =	21. 4 X 6 =

Activity 2

Date: ______ Start: ______ Finish: ______ Score: ______

22. 2 X 8 =	23. 3 X 2 =	24. 5 X 7 =
25. 4 X 9 =	26. 3 X 5 =	27. 5 X 4 =
28. 2 X 2 =	29. 3 X 3 =	30. 5 X 4 =
31. 2 X 9 =	32. 4 X 8 =	33. 3 X 5 =
34. 2 X 7 =	35. 5 X 10 =	36. 4 X 6 =
37. 5 X 10 =	38. 2 X 2 =	39. 3 X 6 =
40. 4 X 3 =	41. 3 X 4 =	42. 4 X 7 =

Activity 1

Date: ______ Start: ______ Finish: ______ Score: ______

1) 2 X 2 =	2) 3 X 10 =	3) 4 X 8 =
4) 5 X 4 =	5) 4 X 6 =	6) 3 X 9 =
7) 5 X 5 =	8) 2 X 7 =	9) 3 X 3 =
10) 2 X 2 =	11) 5 X 4 =	12) 4 X 10 =
13) 2 X 8 =	14) 4 X 6 =	15) 5 X 7 =
16) 3 X 5 =	17) 4 X 9 =	18) 2 X 3 =
19) 3 X 4 =	20) 5 X 8 =	21) 4 X 9 =

Activity 2

Date: ______ Start: ______ Finish: ______ Score: ______

22) 2 X 3 =	23) 5 X 2 =	24) 3 X 7 =
25) 5 X 6 =	26) 2 X 5 =	27) 4 X 10 =
28) 3 X 4 =	29) 5 X 2 =	30) 2 X 5 =
31) 3 X 3 =	32) 4 X 6 =	33) 3 X 9 =
34) 2 X 10 =	35) 5 X 8 =	36) 4 X 7 =
37) 3 X 9 =	38) 2 X 8 =	39) 5 X 4 =
40) 4 X 7 =	41) 2 X 2 =	42) 4 X 3 =

Activity 1

Date: ________ Start: ________ Finish: ________ Score: ________

1	2 X 5 =	2	3 X 2 =	3	5 X 10 =
4	4 X 6 =	5	2 X 3 =	6	3 X 9 =
7	4 X 8 =	8	5 X 7 =	9	4 X 4 =
10	2 X 6 =	11	5 X 4 =	12	3 X 5 =
13	4 X 2 =	14	3 X 9 =	15	2 X 3 =
16	5 X 10 =	17	2 X 7 =	18	5 X 8 =
19	4 X 4 =	20	3 X 5 =	21	4 X 10 =

Activity 2

Date: ________ Start: ________ Finish: ________ Score: ________

22	3 X 7 =	23	2 X 9 =	24	5 X 6 =
25	2 X 2 =	26	5 X 3 =	27	4 X 8 =
28	3 X 7 =	29	2 X 3 =	30	5 X 4 =
31	3 X 5 =	32	4 X 6 =	33	3 X 8 =
34	4 X 10 =	35	5 X 2 =	36	2 X 9 =
37	4 X 4 =	38	3 X 10 =	39	2 X 9 =
40	5 X 5 =	41	2 X 8 =	42	5 X 6 =

Activity 1

Date: ________ Start: ________ Finish: ________ Score: ________

1. 3 X 5 =	2. 2 X 3 =	3. 4 X 7 =
4. 5 X 4 =	5. 3 X 9 =	6. 5 X 6 =
7. 4 X 8 =	8. 2 X 10 =	9. 5 X 2 =
10. 4 X 8 =	11. 2 X 10 =	12. 3 X 2 =
13. 5 X 3 =	14. 4 X 7 =	15. 2 X 9 =
16. 3 X 4 =	17. 2 X 5 =	18. 4 X 6 =
19. 3 X 9 =	20. 5 X 3 =	21. 4 X 7 =

Activity 2

Date: ________ Start: ________ Finish: ________ Score: ________

22. 3 X 8 =	23. 5 X 6 =	24. 2 X 5 =
25. 3 X 4 =	26. 2 X 2 =	27. 4 X 10 =
28. 5 X 4 =	29. 2 X 5 =	30. 3 X 8 =
31. 5 X 3 =	32. 4 X 9 =	33. 5 X 6 =
34. 4 X 10 =	35. 3 X 7 =	36. 2 X 2 =
37. 5 X 6 =	38. 4 X 7 =	39. 3 X 5 =
40. 2 X 3 =	41. 4 X 8 =	42. 3 X 9 =

Practice: Multiplication Table 6

Activity 1

Date: ________ Start: ________ Finish: ________ Score: ________

1. 6 X 2 =	2. 6 X 4 =	3. 6 X 9 =
4. 6 X 3 =	5. 6 X 6 =	6. 6 X 7 =
7. 6 X 8 =	8. 6 X 10 =	9. 6 X 5 =
10. 6 X 3 =	11. 6 X 7 =	12. 6 X 9 =
13. 6 X 8 =	14. 6 X 2 =	15. 6 X 10 =
16. 6 X 5 =	17. 6 X 6 =	18. 6 X 4 =
19. 6 X 3 =	20. 6 X 10 =	21. 6 X 5 =

Activity 2

Date: ________ Start: ________ Finish: ________ Score: ________

22. 6 X 6 =	23. 6 X 2 =	24. 6 X 8 =
25. 6 X 7 =	26. 6 X 4 =	27. 6 X 9 =
28. 6 X 7 =	29. 6 X 10 =	30. 6 X 8 =
31. 6 X 2 =	32. 6 X 5 =	33. 6 X 6 =
34. 6 X 3 =	35. 6 X 4 =	36. 6 X 9 =
37. 6 X 10 =	38. 6 X 2 =	39. 6 X 3 =
40. 6 X 7 =	41. 6 X 6 =	42. 6 X 5 =

Activity 1

Date: ______ Start: ______ Finish: ______ Score: ______

1) 6 X 2 =	2) 6 X 4 =	3) 6 X 8 =
4) 6 X 7 =	5) 6 X 10 =	6) 6 X 5 =
7) 6 X 3 =	8) 6 X 9 =	9) 6 X 6 =
10) 6 X 3 =	11) 6 X 7 =	12) 6 X 4 =
13) 6 X 9 =	14) 6 X 6 =	15) 6 X 5 =
16) 6 X 8 =	17) 6 X 2 =	18) 6 X 10 =
19) 6 X 5 =	20) 6 X 3 =	21) 6 X 10 =

Activity 2

Date: ______ Start: ______ Finish: ______ Score: ______

22) 6 X 8 =	23) 6 X 7 =	24) 6 X 9 =
25) 6 X 4 =	26) 6 X 6 =	27) 6 X 2 =
28) 6 X 9 =	29) 6 X 5 =	30) 6 X 6 =
31) 6 X 2 =	32) 6 X 7 =	33) 6 X 8 =
34) 6 X 10 =	35) 6 X 4 =	36) 6 X 3 =
37) 6 X 5 =	38) 6 X 7 =	39) 6 X 2 =
40) 6 X 10 =	41) 6 X 4 =	42) 6 X 8 =

Practice: Multiplication Table 6

Activity 1

Date: ______ Start: ______ Finish: ______ Score: ______

1. 6 X 2 =	2. 6 X 3 =	3. 6 X 7 =
4. 6 X 5 =	5. 6 X 9 =	6. 6 X 6 =
7. 6 X 4 =	8. 6 X 8 =	9. 6 X 10 =
10. 6 X 6 =	11. 6 X 8 =	12. 6 X 7 =
13. 6 X 10 =	14. 6 X 4 =	15. 6 X 5 =
16. 6 X 3 =	17. 6 X 2 =	18. 6 X 9 =
19. 6 X 8 =	20. 6 X 2 =	21. 6 X 7 =

Activity 2

Date: ______ Start: ______ Finish: ______ Score: ______

22. 6 X 10 =	23. 6 X 5 =	24. 6 X 3 =
25. 6 X 4 =	26. 6 X 9 =	27. 6 X 6 =
28. 6 X 5 =	29. 6 X 7 =	30. 6 X 4 =
31. 6 X 6 =	32. 6 X 8 =	33. 6 X 9 =
34. 6 X 2 =	35. 6 X 10 =	36. 6 X 3 =
37. 6 X 2 =	38. 6 X 3 =	39. 6 X 6 =
40. 6 X 7 =	41. 6 X 10 =	42. 6 X 4 =

Activity 1

Date: ______ Start: ______ Finish: ______ Score: ______

1) 7 X 3 =	2) 7 X 7 =	3) 7 X 5 =
4) 7 X 9 =	5) 7 X 8 =	6) 7 X 6 =
7) 7 X 2 =	8) 7 X 10 =	9) 7 X 4 =
10) 7 X 5 =	11) 7 X 7 =	12) 7 X 10 =
13) 7 X 9 =	14) 7 X 6 =	15) 7 X 2 =
16) 7 X 8 =	17) 7 X 3 =	18) 7 X 4 =
19) 7 X 2 =	20) 7 X 10 =	21) 7 X 6 =

Activity 2

Date: ______ Start: ______ Finish: ______ Score: ______

22) 7 X 7 =	23) 7 X 9 =	24) 7 X 3 =
25) 7 X 4 =	26) 7 X 5 =	27) 7 X 8 =
28) 7 X 5 =	29) 7 X 10 =	30) 7 X 2 =
31) 7 X 9 =	32) 7 X 4 =	33) 7 X 3 =
34) 7 X 6 =	35) 7 X 7 =	36) 7 X 8 =
37) 7 X 4 =	38) 7 X 9 =	39) 7 X 7 =
40) 7 X 6 =	41) 7 X 5 =	42) 7 X 3 =

Activity 1

Date: ________ Start: ________ Finish: ________ Score: ________

1	7 X 6 =	2	7 X 8 =	3	7 X 9 =
4	7 X 7 =	5	7 X 3 =	6	7 X 5 =
7	7 X 2 =	8	7 X 4 =	9	7 X 10 =
10	7 X 8 =	11	7 X 3 =	12	7 X 6 =
13	7 X 10 =	14	7 X 4 =	15	7 X 9 =
16	7 X 5 =	17	7 X 2 =	18	7 X 7 =
19	7 X 3 =	20	7 X 6 =	21	7 X 7 =

Activity 2

Date: ________ Start: ________ Finish: ________ Score: ________

22	7 X 4 =	23	7 X 8 =	24	7 X 9 =
25	7 X 2 =	26	7 X 5 =	27	7 X 10 =
28	7 X 7 =	29	7 X 10 =	30	7 X 9 =
31	7 X 3 =	32	7 X 2 =	33	7 X 6 =
34	7 X 5 =	35	7 X 8 =	36	7 X 4 =
37	7 X 5 =	38	7 X 3 =	39	7 X 2 =
40	7 X 10 =	41	7 X 8 =	42	7 X 7 =

Activity 1

Date: ______ Start: ______ Finish: ______ Score: ______

1 7 X 10 =	2 7 X 6 =	3 7 X 8 =
4 7 X 5 =	5 7 X 7 =	6 7 X 2 =
7 7 X 4 =	8 7 X 9 =	9 7 X 3 =
10 7 X 9 =	11 7 X 10 =	12 7 X 6 =
13 7 X 7 =	14 7 X 2 =	15 7 X 4 =
16 7 X 8 =	17 7 X 3 =	18 7 X 5 =
19 7 X 10 =	20 7 X 4 =	21 7 X 6 =

Activity 2

Date: ______ Start: ______ Finish: ______ Score: ______

22 7 X 3 =	23 7 X 2 =	24 7 X 5 =
25 7 X 9 =	26 7 X 7 =	27 7 X 8 =
28 7 X 3 =	29 7 X 4 =	30 7 X 9 =
31 7 X 8 =	32 7 X 7 =	33 7 X 2 =
34 7 X 6 =	35 7 X 5 =	36 7 X 10 =
37 7 X 4 =	38 7 X 10 =	39 7 X 8 =
40 7 X 5 =	41 7 X 6 =	42 7 X 3 =

Activity 1

Date: Start: Finish: Score:

1 8 X 2 =	2 8 X 6 =	3 8 X 5 =
4 8 X 10 =	5 8 X 4 =	6 8 X 8 =
7 8 X 3 =	8 8 X 7 =	9 8 X 9 =
10 8 X 7 =	11 8 X 9 =	12 8 X 6 =
13 8 X 10 =	14 8 X 4 =	15 8 X 3 =
16 8 X 5 =	17 8 X 8 =	18 8 X 2 =
19 8 X 6 =	20 8 X 4 =	21 8 X 9 =

Activity 2

Date: Start: Finish: Score:

22 8 X 8 =	23 8 X 7 =	24 8 X 10 =
25 8 X 5 =	26 8 X 2 =	27 8 X 3 =
28 8 X 7 =	29 8 X 5 =	30 8 X 9 =
31 8 X 6 =	32 8 X 2 =	33 8 X 3 =
34 8 X 8 =	35 8 X 10 =	36 8 X 4 =
37 8 X 3 =	38 8 X 7 =	39 8 X 6 =
40 8 X 10 =	41 8 X 5 =	42 8 X 8 =

Activity 1

Date: ______ Start: ______ Finish: ______ Score: ______

1. 8 X 4 =	2. 8 X 7 =	3. 8 X 9 =
4. 8 X 2 =	5. 8 X 8 =	6. 8 X 3 =
7. 8 X 10 =	8. 8 X 5 =	9. 8 X 6 =
10. 8 X 8 =	11. 8 X 6 =	12. 8 X 3 =
13. 8 X 7 =	14. 8 X 10 =	15. 8 X 9 =
16. 8 X 5 =	17. 8 X 4 =	18. 8 X 2 =
19. 8 X 4 =	20. 8 X 5 =	21. 8 X 3 =

Activity 2

Date: ______ Start: ______ Finish: ______ Score: ______

22. 8 X 9 =	23. 8 X 7 =	24. 8 X 2 =
25. 8 X 10 =	26. 8 X 8 =	27. 8 X 6 =
28. 8 X 2 =	29. 8 X 6 =	30. 8 X 10 =
31. 8 X 4 =	32. 8 X 9 =	33. 8 X 5 =
34. 8 X 7 =	35. 8 X 8 =	36. 8 X 3 =
37. 8 X 5 =	38. 8 X 3 =	39. 8 X 7 =
40. 8 X 4 =	41. 8 X 10 =	42. 8 X 6 =

Practice: Multiplication Table 8

Activity 1

Date: ______ Start: ______ Finish: ______ Score: ______

1. 8 X 6 =	2. 8 X 4 =	3. 8 X 8 =
4. 8 X 9 =	5. 8 X 3 =	6. 8 X 10 =
7. 8 X 5 =	8. 8 X 2 =	9. 8 X 7 =
10. 8 X 2 =	11. 8 X 8 =	12. 8 X 7 =
13. 8 X 9 =	14. 8 X 6 =	15. 8 X 4 =
16. 8 X 5 =	17. 8 X 10 =	18. 8 X 3 =
19. 8 X 5 =	20. 8 X 3 =	21. 8 X 8 =

Activity 2

Date: ______ Start: ______ Finish: ______ Score: ______

22. 8 X 7 =	23. 8 X 9 =	24. 8 X 10 =
25. 8 X 2 =	26. 8 X 4 =	27. 8 X 6 =
28. 8 X 7 =	29. 8 X 2 =	30. 8 X 5 =
31. 8 X 3 =	32. 8 X 10 =	33. 8 X 9 =
34. 8 X 6 =	35. 8 X 8 =	36. 8 X 4 =
37. 8 X 10 =	38. 8 X 8 =	39. 8 X 5 =
40. 8 X 7 =	41. 8 X 9 =	42. 8 X 6 =

Activity 1

Date: ______ Start: ______ Finish: ______ Score: ______

1. 9 X 6 =	2. 9 X 9 =	3. 9 X 8 =
4. 9 X 10 =	5. 9 X 4 =	6. 9 X 3 =
7. 9 X 7 =	8. 9 X 2 =	9. 9 X 5 =
10. 9 X 10 =	11. 9 X 5 =	12. 9 X 3 =
13. 9 X 6 =	14. 9 X 2 =	15. 9 X 7 =
16. 9 X 9 =	17. 9 X 4 =	18. 9 X 8 =
19. 9 X 6 =	20. 9 X 3 =	21. 9 X 4 =

Activity 2

Date: ______ Start: ______ Finish: ______ Score: ______

22. 9 X 10 =	23. 9 X 5 =	24. 9 X 2 =
25. 9 X 9 =	26. 9 X 8 =	27. 9 X 7 =
28. 9 X 6 =	29. 9 X 5 =	30. 9 X 10 =
31. 9 X 4 =	32. 9 X 2 =	33. 9 X 9 =
34. 9 X 3 =	35. 9 X 7 =	36. 9 X 8 =
37. 9 X 10 =	38. 9 X 2 =	39. 9 X 9 =
40. 9 X 7 =	41. 9 X 6 =	42. 9 X 8 =

Practice: Multiplication Table 9

Activity 1

Date: ________ Start: ________ Finish: ________ Score: ________

1. 9 X 8 =	2. 9 X 5 =	3. 9 X 2 =
4. 9 X 7 =	5. 9 X 3 =	6. 9 X 10 =
7. 9 X 4 =	8. 9 X 6 =	9. 9 X 9 =
10. 9 X 10 =	11. 9 X 7 =	12. 9 X 2 =
13. 9 X 6 =	14. 9 X 9 =	15. 9 X 8 =
16. 9 X 4 =	17. 9 X 3 =	18. 9 X 5 =
19. 9 X 4 =	20. 9 X 7 =	21. 9 X 2 =

Activity 2

Date: ________ Start: ________ Finish: ________ Score: ________

22. 9 X 5 =	23. 9 X 8 =	24. 9 X 3 =
25. 9 X 6 =	26. 9 X 10 =	27. 9 X 9 =
28. 9 X 5 =	29. 9 X 6 =	30. 9 X 2 =
31. 9 X 10 =	32. 9 X 8 =	33. 9 X 4 =
34. 9 X 3 =	35. 9 X 7 =	36. 9 X 9 =
37. 9 X 2 =	38. 9 X 4 =	39. 9 X 8 =
40. 9 X 10 =	41. 9 X 6 =	42. 9 X 3 =

Activity 1

Date: ________ Start: ________ Finish: ________ Score: ________

1. 9 X 10 =	2. 9 X 9 =	3. 9 X 6 =
4. 9 X 5 =	5. 9 X 3 =	6. 9 X 2 =
7. 9 X 4 =	8. 9 X 7 =	9. 9 X 8 =
10. 9 X 6 =	11. 9 X 7 =	12. 9 X 8 =
13. 9 X 9 =	14. 9 X 3 =	15. 9 X 2 =
16. 9 X 4 =	17. 9 X 10 =	18. 9 X 5 =
19. 9 X 2 =	20. 9 X 5 =	21. 9 X 3 =

Activity 2

Date: ________ Start: ________ Finish: ________ Score: ________

22. 9 X 4 =	23. 9 X 10 =	24. 9 X 6 =
25. 9 X 7 =	26. 9 X 8 =	27. 9 X 9 =
28. 9 X 3 =	29. 9 X 7 =	30. 9 X 10 =
31. 9 X 9 =	32. 9 X 4 =	33. 9 X 8 =
34. 9 X 6 =	35. 9 X 5 =	36. 9 X 2 =
37. 9 X 8 =	38. 9 X 5 =	39. 9 X 4 =
40. 9 X 2 =	41. 9 X 10 =	42. 9 X 7 =

Activity 1

Date: ______ Start: ______ Finish: ______ Score: ______

1. 10 X 3 = ☐	2. 10 X 9 = ☐	3. 10 X 10 = ☐
4. 10 X 8 = ☐	5. 10 X 4 = ☐	6. 10 X 6 = ☐
7. 10 X 7 = ☐	8. 10 X 5 = ☐	9. 10 X 2 = ☐
10. 10 X 10 = ☐	11. 10 X 7 = ☐	12. 10 X 3 = ☐
13. 10 X 6 = ☐	14. 10 X 8 = ☐	15. 10 X 5 = ☐
16. 10 X 2 = ☐	17. 10 X 9 = ☐	18. 10 X 4 = ☐
19. 10 X 9 = ☐	20. 10 X 6 = ☐	21. 10 X 2 = ☐

Activity 2

Date: ______ Start: ______ Finish: ______ Score: ______

22. 10 X 3 = ☐	23. 10 X 7 = ☐	24. 10 X 4 = ☐
25. 10 X 10 = ☐	26. 10 X 5 = ☐	27. 10 X 8 = ☐
28. 10 X 6 = ☐	29. 10 X 10 = ☐	30. 10 X 2 = ☐
31. 10 X 8 = ☐	32. 10 X 3 = ☐	33. 10 X 4 = ☐
34. 10 X 5 = ☐	35. 10 X 7 = ☐	36. 10 X 9 = ☐
37. 10 X 4 = ☐	38. 10 X 7 = ☐	39. 10 X 3 = ☐
40. 10 X 6 = ☐	41. 10 X 8 = ☐	42. 10 X 5 = ☐

Activity 1

Date: ______ Start: ______ Finish: ______ Score: ______

1. 10 X 6 =	2. 10 X 7 =	3. 10 X 10 =
4. 10 X 4 =	5. 10 X 3 =	6. 10 X 2 =
7. 10 X 9 =	8. 10 X 5 =	9. 10 X 8 =
10. 10 X 9 =	11. 10 X 5 =	12. 10 X 6 =
13. 10 X 3 =	14. 10 X 2 =	15. 10 X 7 =
16. 10 X 4 =	17. 10 X 8 =	18. 10 X 10 =
19. 10 X 9 =	20. 10 X 2 =	21. 10 X 4 =

Activity 2

Date: ______ Start: ______ Finish: ______ Score: ______

22. 10 X 8 =	23. 10 X 5 =	24. 10 X 7 =
25. 10 X 10 =	26. 10 X 6 =	27. 10 X 3 =
28. 10 X 8 =	29. 10 X 5 =	30. 10 X 6 =
31. 10 X 7 =	32. 10 X 2 =	33. 10 X 4 =
34. 10 X 10 =	35. 10 X 3 =	36. 10 X 9 =
37. 10 X 8 =	38. 10 X 9 =	39. 10 X 10 =
40. 10 X 3 =	41. 10 X 6 =	42. 10 X 5 =

Practice: Multiplication Table 10

Activity 1

Date: ________ Start: ________ Finish: ________ Score: ________

1 10 X 10 =	2 10 X 5 =	3 10 X 8 =
4 10 X 9 =	5 10 X 4 =	6 10 X 6 =
7 10 X 3 =	8 10 X 2 =	9 10 X 7 =
10 10 X 9 =	11 10 X 3 =	12 10 X 6 =
13 10 X 5 =	14 10 X 7 =	15 10 X 8 =
16 10 X 2 =	17 10 X 4 =	18 10 X 10 =
19 10 X 3 =	20 10 X 6 =	21 10 X 4 =

Activity 2

Date: ________ Start: ________ Finish: ________ Score: ________

22 10 X 10 =	23 10 X 2 =	24 10 X 9 =
25 10 X 7 =	26 10 X 5 =	27 10 X 8 =
28 10 X 7 =	29 10 X 6 =	30 10 X 8 =
31 10 X 9 =	32 10 X 4 =	33 10 X 2 =
34 10 X 3 =	35 10 X 10 =	36 10 X 5 =
37 10 X 3 =	38 10 X 6 =	39 10 X 4 =
40 10 X 5 =	41 10 X 10 =	42 10 X 2 =

Activity 1

Date: ______ Start: ______ Finish: ______ Score: ______

1. 8 X 8 =	2. 10 X 5 =	3. 6 X 9 =
4. 7 X 10 =	5. 9 X 7 =	6. 8 X 2 =
7. 9 X 3 =	8. 7 X 4 =	9. 10 X 6 =
10. 6 X 5 =	11. 7 X 8 =	12. 9 X 3 =
13. 10 X 9 =	14. 6 X 10 =	15. 8 X 7 =
16. 6 X 6 =	17. 7 X 4 =	18. 8 X 2 =
19. 9 X 4 =	20. 10 X 9 =	21. 6 X 3 =

Activity 2

Date: ______ Start: ______ Finish: ______ Score: ______

22. 10 X 5 =	23. 7 X 8 =	24. 9 X 7 =
25. 8 X 6 =	26. 6 X 2 =	27. 7 X 10 =
28. 8 X 4 =	29. 9 X 10 =	30. 10 X 2 =
31. 7 X 7 =	32. 8 X 9 =	33. 6 X 8 =
34. 10 X 6 =	35. 9 X 5 =	36. 8 X 3 =
37. 10 X 5 =	38. 6 X 2 =	39. 9 X 9 =
40. 7 X 4 =	41. 10 X 6 =	42. 6 X 10 =

Activity 1

Date: ________ Start: ________ Finish: ________ Score: ________

1	6 X 3 =	2	9 X 2 =	3	8 X 7 =
4	7 X 10 =	5	10 X 8 =	6	9 X 9 =
7	10 X 5 =	8	7 X 6 =	9	8 X 4 =
10	6 X 3 =	11	8 X 5 =	12	10 X 8 =
13	7 X 6 =	14	6 X 2 =	15	9 X 7 =
16	10 X 9 =	17	6 X 10 =	18	7 X 4 =
19	8 X 8 =	20	9 X 2 =	21	10 X 5 =

Activity 2

Date: ________ Start: ________ Finish: ________ Score: ________

22	8 X 3 =	23	7 X 6 =	24	6 X 10 =
25	9 X 7 =	26	10 X 9 =	27	8 X 4 =
28	9 X 7 =	29	6 X 6 =	30	7 X 4 =
31	10 X 2 =	32	7 X 8 =	33	9 X 5 =
34	6 X 3 =	35	8 X 10 =	36	9 X 9 =
37	10 X 2 =	38	7 X 4 =	39	8 X 5 =
40	6 X 10 =	41	10 X 6 =	42	6 X 9 =

Activity 1

Date: ________ Start: ________ Finish: ________ Score: ________

1. 7 X 10 =	2. 9 X 3 =	3. 6 X 5 =
4. 10 X 9 =	5. 8 X 6 =	6. 7 X 7 =
7. 9 X 8 =	8. 8 X 4 =	9. 10 X 2 =
10. 6 X 6 =	11. 9 X 2 =	12. 10 X 3 =
13. 7 X 8 =	14. 8 X 10 =	15. 6 X 9 =
16. 8 X 4 =	17. 7 X 7 =	18. 9 X 5 =
19. 10 X 3 =	20. 6 X 7 =	21. 10 X 8 =

Activity 2

Date: ________ Start: ________ Finish: ________ Score: ________

22. 6 X 4 =	23. 8 X 10 =	24. 9 X 5 =
25. 7 X 6 =	26. 9 X 9 =	27. 7 X 2 =
28. 10 X 5 =	29. 8 X 4 =	30. 6 X 2 =
31. 7 X 10 =	32. 8 X 3 =	33. 10 X 6 =
34. 9 X 8 =	35. 6 X 7 =	36. 8 X 9 =
37. 9 X 7 =	38. 6 X 9 =	39. 7 X 8 =
40. 10 X 4 =	41. 9 X 5 =	42. 7 X 10 =

Activity 1

Date: ______ Start: ______ Finish: ______ Score: ______

1	8 X 4 =	2	6 X 8 =	3	10 X 5 =
4	9 X 6 =	5	7 X 10 =	6	10 X 7 =
7	6 X 9 =	8	8 X 2 =	9	7 X 3 =
10	9 X 10 =	11	8 X 5 =	12	10 X 3 =
13	7 X 9 =	14	6 X 6 =	15	9 X 7 =
16	6 X 4 =	17	9 X 8 =	18	10 X 2 =
19	7 X 9 =	20	8 X 5 =	21	6 X 6 =

Activity 2

Date: ______ Start: ______ Finish: ______ Score: ______

22	8 X 8 =	23	9 X 7 =	24	7 X 10 =
25	10 X 4 =	26	9 X 2 =	27	10 X 3 =
28	8 X 8 =	29	7 X 9 =	30	6 X 3 =
31	9 X 5 =	32	10 X 2 =	33	7 X 7 =
34	8 X 6 =	35	6 X 4 =	36	7 X 10 =
37	8 X 6 =	38	10 X 9 =	39	9 X 7 =
40	6 X 8 =	41	9 X 3 =	42	7 X 2 =

Activity 1

Date: ________ Start: ________ Finish: ________ Score: ________

1. 8 X 5 =	2. 7 X 8 =	3. 9 X 2 =
4. 6 X 4 =	5. 10 X 9 =	6. 8 X 3 =
7. 6 X 10 =	8. 10 X 6 =	9. 7 X 7 =
10. 9 X 9 =	11. 6 X 5 =	12. 9 X 2 =
13. 7 X 3 =	14. 10 X 10 =	15. 8 X 7 =
16. 10 X 4 =	17. 8 X 8 =	18. 7 X 6 =
19. 6 X 8 =	20. 9 X 4 =	21. 7 X 3 =

Activity 2

Date: ________ Start: ________ Finish: ________ Score: ________

22. 8 X 7 =	23. 6 X 9 =	24. 9 X 6 =
25. 10 X 10 =	26. 9 X 5 =	27. 10 X 2 =
28. 8 X 9 =	29. 6 X 2 =	30. 7 X 5 =
31. 8 X 10 =	32. 10 X 3 =	33. 9 X 7 =
34. 7 X 4 =	35. 6 X 8 =	36. 8 X 6 =
37. 9 X 5 =	38. 7 X 2 =	39. 6 X 10 =
40. 10 X 7 =	41. 9 X 4 =	42. 8 X 8 =

Practice: Multiplication Table 11

Activity 1

Date: ________ Start: ________ Finish: ________ Score: ________

1. 11 X 4 =	2. 11 X 2 =	3. 11 X 5 =
4. 11 X 9 =	5. 11 X 8 =	6. 11 X 3 =
7. 11 X 6 =	8. 11 X 7 =	9. 11 X 10 =
10. 11 X 3 =	11. 11 X 7 =	12. 11 X 8 =
13. 11 X 4 =	14. 11 X 5 =	15. 11 X 6 =
16. 11 X 2 =	17. 11 X 9 =	18. 11 X 10 =
19. 11 X 8 =	20. 11 X 10 =	21. 11 X 5 =

Activity 2

Date: ________ Start: ________ Finish: ________ Score: ________

22. 11 X 4 =	23. 11 X 6 =	24. 11 X 2 =
25. 11 X 9 =	26. 11 X 3 =	27. 11 X 7 =
28. 11 X 6 =	29. 11 X 3 =	30. 11 X 7 =
31. 11 X 10 =	32. 11 X 4 =	33. 11 X 5 =
34. 11 X 9 =	35. 11 X 2 =	36. 11 X 8 =
37. 11 X 3 =	38. 11 X 4 =	39. 11 X 10 =
40. 11 X 2 =	41. 11 X 8 =	42. 11 X 5 =

Activity 1

Date: ______ Start: ______ Finish: ______ Score: ______

1) 11 X 5 =	2) 11 X 8 =	3) 11 X 9 =
4) 11 X 3 =	5) 11 X 2 =	6) 11 X 4 =
7) 11 X 7 =	8) 11 X 6 =	9) 11 X 10 =
10) 11 X 5 =	11) 11 X 3 =	12) 11 X 4 =
13) 11 X 10 =	14) 11 X 8 =	15) 11 X 7 =
16) 11 X 9 =	17) 11 X 6 =	18) 11 X 2 =
19) 11 X 5 =	20) 11 X 3 =	21) 11 X 4 =

Activity 2

Date: ______ Start: ______ Finish: ______ Score: ______

22) 11 X 9 =	23) 11 X 2 =	24) 11 X 8 =
25) 11 X 10 =	26) 11 X 6 =	27) 11 X 7 =
28) 11 X 5 =	29) 11 X 10 =	30) 11 X 8 =
31) 11 X 2 =	32) 11 X 9 =	33) 11 X 3 =
34) 11 X 4 =	35) 11 X 7 =	36) 11 X 6 =
37) 11 X 2 =	38) 11 X 10 =	39) 11 X 8 =
40) 11 X 9 =	41) 11 X 5 =	42) 11 X 4 =

Practice: Multiplication Table 11

Activity 1

Date: ________ Start: ________ Finish: ________ Score: ________

1. 11 X 4 = ☐
2. 11 X 8 = ☐
3. 11 X 2 = ☐
4. 11 X 5 = ☐
5. 11 X 10 = ☐
6. 11 X 6 = ☐
7. 11 X 9 = ☐
8. 11 X 7 = ☐
9. 11 X 3 = ☐
10. 11 X 5 = ☐
11. 11 X 6 = ☐
12. 11 X 4 = ☐
13. 11 X 9 = ☐
14. 11 X 2 = ☐
15. 11 X 8 = ☐
16. 11 X 7 = ☐
17. 11 X 3 = ☐
18. 11 X 10 = ☐
19. 11 X 6 = ☐
20. 11 X 5 = ☐
21. 11 X 9 = ☐

Activity 2

Date: ________ Start: ________ Finish: ________ Score: ________

22. 11 X 7 = ☐
23. 11 X 4 = ☐
24. 11 X 2 = ☐
25. 11 X 8 = ☐
26. 11 X 3 = ☐
27. 11 X 10 = ☐
28. 11 X 2 = ☐
29. 11 X 5 = ☐
30. 11 X 9 = ☐
31. 11 X 7 = ☐
32. 11 X 3 = ☐
33. 11 X 4 = ☐
34. 11 X 8 = ☐
35. 11 X 6 = ☐
36. 11 X 10 = ☐
37. 11 X 7 = ☐
38. 11 X 10 = ☐
39. 11 X 6 = ☐
40. 11 X 9 = ☐
41. 11 X 2 = ☐
42. 11 X 8 = ☐

Activity 1

Date: ________ Start: ________ Finish: ________ Score: ________

1. 12 X 3 =	2. 12 X 7 =	3. 12 X 10 =
4. 12 X 2 =	5. 12 X 6 =	6. 12 X 9 =
7. 12 X 4 =	8. 12 X 8 =	9. 12 X 5 =
10. 12 X 10 =	11. 12 X 5 =	12. 12 X 6 =
13. 12 X 8 =	14. 12 X 7 =	15. 12 X 3 =
16. 12 X 9 =	17. 12 X 4 =	18. 12 X 2 =
19. 12 X 9 =	20. 12 X 4 =	21. 12 X 8 =

Activity 2

Date: ________ Start: ________ Finish: ________ Score: ________

22. 12 X 3 =	23. 12 X 6 =	24. 12 X 5 =
25. 12 X 7 =	26. 12 X 2 =	27. 12 X 10 =
28. 12 X 2 =	29. 12 X 6 =	30. 12 X 7 =
31. 12 X 5 =	32. 12 X 9 =	33. 12 X 10 =
34. 12 X 3 =	35. 12 X 4 =	36. 12 X 8 =
37. 12 X 3 =	38. 12 X 4 =	39. 12 X 9 =
40. 12 X 8 =	41. 12 X 2 =	42. 12 X 5 =

Practice: Multiplication Table 12

Activity 1

Date: ________ Start: ________ Finish: ________ Score: ________

1) 12 X 8 =	2) 12 X 6 =	3) 12 X 3 =
4) 12 X 4 =	5) 12 X 10 =	6) 12 X 5 =
7) 12 X 2 =	8) 12 X 7 =	9) 12 X 9 =
10) 12 X 6 =	11) 12 X 3 =	12) 12 X 9 =
13) 12 X 2 =	14) 12 X 10 =	15) 12 X 8 =
16) 12 X 4 =	17) 12 X 7 =	18) 12 X 5 =
19) 12 X 2 =	20) 12 X 3 =	21) 12 X 9 =

Activity 2

Date: ________ Start: ________ Finish: ________ Score: ________

22) 12 X 5 =	23) 12 X 6 =	24) 12 X 7 =
25) 12 X 10 =	26) 12 X 8 =	27) 12 X 4 =
28) 12 X 9 =	29) 12 X 2 =	30) 12 X 7 =
31) 12 X 3 =	32) 12 X 4 =	33) 12 X 5 =
34) 12 X 10 =	35) 12 X 6 =	36) 12 X 8 =
37) 12 X 3 =	38) 12 X 6 =	39) 12 X 8 =
40) 12 X 10 =	41) 12 X 7 =	42) 12 X 5 =

Activity 1

Date: Start: Finish: Score:

1 $12 \times 4 =$	2 $12 \times 2 =$	3 $12 \times 8 =$
4 $12 \times 7 =$	5 $12 \times 9 =$	6 $12 \times 10 =$
7 $12 \times 5 =$	8 $12 \times 3 =$	9 $12 \times 6 =$
10 $12 \times 9 =$	11 $12 \times 4 =$	12 $12 \times 5 =$
13 $12 \times 8 =$	14 $12 \times 2 =$	15 $12 \times 3 =$
16 $12 \times 7 =$	17 $12 \times 10 =$	18 $12 \times 6 =$
19 $12 \times 9 =$	20 $12 \times 4 =$	21 $12 \times 2 =$

Activity 2

Date: Start: Finish: Score:

22 $12 \times 10 =$	23 $12 \times 3 =$	24 $12 \times 7 =$
25 $12 \times 5 =$	26 $12 \times 8 =$	27 $12 \times 6 =$
28 $12 \times 4 =$	29 $12 \times 3 =$	30 $12 \times 9 =$
31 $12 \times 5 =$	32 $12 \times 6 =$	33 $12 \times 2 =$
34 $12 \times 10 =$	35 $12 \times 7 =$	36 $12 \times 8 =$
37 $12 \times 6 =$	38 $12 \times 3 =$	39 $12 \times 10 =$
40 $12 \times 8 =$	41 $12 \times 4 =$	42 $12 \times 5 =$

Review: Multiplication Table 2 to 12 Mixed

Activity 1

Date: ______ Start: ______ Finish: ______ Score: ______

1. 6 X 5 =
2. 8 X 4 =
3. 5 X 8 =
4. 12 X 3 =
5. 4 X 6 =
6. 2 X 7 =
7. 7 X 9 =
8. 9 X 10 =
9. 3 X 2 =
10. 10 X 3 =
11. 11 X 10 =
12. 3 X 9 =
13. 10 X 4 =
14. 6 X 8 =
15. 2 X 5 =
16. 8 X 6 =
17. 4 X 7 =
18. 7 X 2 =
19. 9 X 3 =
20. 5 X 2 =
21. 11 X 10 =

Activity 2

Date: ______ Start: ______ Finish: ______ Score: ______

22. 12 X 6 =
23. 4 X 9 =
24. 6 X 7 =
25. 10 X 4 =
26. 2 X 8 =
27. 9 X 5 =
28. 7 X 10 =
29. 5 X 8 =
30. 12 X 5 =
31. 8 X 3 =
32. 11 X 9 =
33. 3 X 4 =
34. 11 X 6 =
35. 2 X 2 =
36. 4 X 7 =
37. 9 X 2 =
38. 10 X 9 =
39. 3 X 6 =
40. 12 X 10 =
41. 6 X 8 =
42. 8 X 7 =

Activity 1

Date: ______ Start: ______ Finish: ______ Score: ______

1	7 X 9 =	2	9 X 10 =	3	6 X 6 =
4	8 X 5 =	5	10 X 8 =	6	5 X 3 =
7	12 X 4 =	8	3 X 7 =	9	2 X 2 =
10	11 X 9 =	11	4 X 8 =	12	5 X 6 =
13	8 X 4 =	14	4 X 10 =	15	12 X 3 =
16	3 X 7 =	17	7 X 2 =	18	6 X 5 =
19	10 X 7 =	20	9 X 8 =	21	11 X 6 =

Activity 2

Date: ______ Start: ______ Finish: ______ Score: ______

22	2 X 3 =	23	9 X 2 =	24	8 X 4 =
25	6 X 5 =	26	4 X 9 =	27	3 X 10 =
28	7 X 3 =	29	12 X 10 =	30	2 X 2 =
31	5 X 4 =	32	10 X 7 =	33	11 X 9 =
34	3 X 8 =	35	10 X 5 =	36	8 X 6 =
37	4 X 8 =	38	11 X 9 =	39	12 X 4 =
40	6 X 10 =	41	2 X 3 =	42	5 X 7 =

Activity 1

Date: ____ Start: ____ Finish: ____ Score: ____

1. 4 X 5 =
2. 12 X 7 =
3. 3 X 8 =
4. 5 X 3 =
5. 11 X 2 =
6. 2 X 9 =
7. 9 X 4 =
8. 7 X 10 =
9. 8 X 6 =
10. 10 X 10 =
11. 6 X 5 =
12. 8 X 8 =
13. 9 X 7 =
14. 10 X 6 =
15. 6 X 9 =
16. 7 X 2 =
17. 5 X 3 =
18. 2 X 4 =
19. 12 X 2 =
20. 11 X 5 =
21. 3 X 9 =

Activity 2

Date: ____ Start: ____ Finish: ____ Score: ____

22. 4 X 10 =
23. 11 X 4 =
24. 9 X 7 =
25. 5 X 6 =
26. 10 X 3 =
27. 7 X 8 =
28. 3 X 5 =
29. 8 X 7 =
30. 6 X 4 =
31. 2 X 6 =
32. 4 X 10 =
33. 12 X 8 =
34. 3 X 3 =
35. 2 X 9 =
36. 4 X 2 =
37. 12 X 3 =
38. 9 X 2 =
39. 5 X 7 =
40. 6 X 8 =
41. 7 X 5 =
42. 8 X 4 =

Activity 1

Date: ______ Start: ______ Finish: ______ Score: ______

1. 4 X 4 =
2. 2 X 6 =
3. 3 X 9 =
4. 12 X 7 =
5. 9 X 2 =
6. 8 X 10 =
7. 5 X 3 =
8. 10 X 5 =
9. 7 X 8 =
10. 11 X 2 =
11. 6 X 9 =
12. 5 X 6 =
13. 9 X 4 =
14. 11 X 3 =
15. 10 X 7 =
16. 3 X 8 =
17. 6 X 5 =
18. 2 X 10 =
19. 12 X 4 =
20. 7 X 7 =
21. 8 X 6 =

Activity 2

Date: ______ Start: ______ Finish: ______ Score: ______

22. 4 X 2 =
23. 5 X 8 =
24. 7 X 10 =
25. 8 X 9 =
26. 10 X 3 =
27. 3 X 5 =
28. 9 X 8 =
29. 12 X 6 =
30. 4 X 9 =
31. 2 X 4 =
32. 6 X 2 =
33. 11 X 5 =
34. 6 X 10 =
35. 12 X 7 =
36. 5 X 3 =
37. 11 X 4 =
38. 9 X 7 =
39. 8 X 3 =
40. 10 X 8 =
41. 2 X 5 =
42. 4 X 10 =

Activity 1

Date: ________ Start: ________ Finish: ________ Score: ________

1	3 X 9 =		2	4 X 3 =		3	12 X 2 =	
4	6 X 4 =		5	9 X 8 =		6	11 X 5 =	
7	2 X 6 =		8	5 X 7 =		9	10 X 10 =	
10	8 X 4 =		11	7 X 7 =		12	3 X 2 =	
13	6 X 3 =		14	2 X 10 =		15	7 X 9 =	
16	11 X 8 =		17	4 X 5 =		18	8 X 6 =	
19	12 X 4 =		20	10 X 3 =		21	5 X 7 =	

Activity 2

Date: ________ Start: ________ Finish: ________ Score: ________

22	9 X 10 =		23	5 X 5 =		24	12 X 2 =	
25	7 X 9 =		26	11 X 8 =		27	6 X 6 =	
28	10 X 3 =		29	8 X 7 =		30	3 X 5 =	
31	2 X 8 =		32	9 X 2 =		33	4 X 9 =	
34	9 X 10 =		35	12 X 4 =		36	6 X 6 =	
37	2 X 7 =		38	7 X 2 =		39	5 X 10 =	
40	8 X 9 =		41	3 X 6 =		42	11 X 3 =	

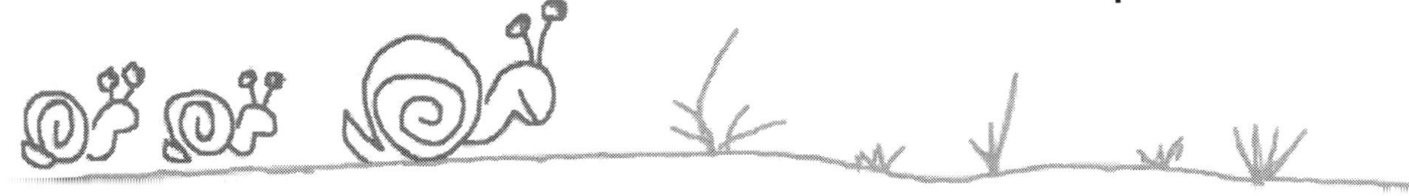

Activity 1

Date: ______ Start: ______ Finish: ______ Score: ______

1. 13 X 4 =	2. 13 X 5 =	3. 13 X 2 =
4. 13 X 10 =	5. 13 X 7 =	6. 13 X 6 =
7. 13 X 9 =	8. 13 X 3 =	9. 13 X 8 =
10. 13 X 10 =	11. 13 X 3 =	12. 13 X 4 =
13. 13 X 2 =	14. 13 X 5 =	15. 13 X 7 =
16. 13 X 9 =	17. 13 X 8 =	18. 13 X 6 =
19. 13 X 2 =	20. 13 X 4 =	21. 13 X 3 =

Activity 2

Date: ______ Start: ______ Finish: ______ Score: ______

22. 13 X 10 =	23. 13 X 5 =	24. 13 X 9 =
25. 13 X 8 =	26. 13 X 7 =	27. 13 X 6 =
28. 13 X 4 =	29. 13 X 8 =	30. 13 X 10 =
31. 13 X 5 =	32. 13 X 7 =	33. 13 X 2 =
34. 13 X 6 =	35. 13 X 9 =	36. 13 X 3 =
37. 13 X 8 =	38. 13 X 3 =	39. 13 X 5 =
40. 13 X 2 =	41. 13 X 6 =	42. 13 X 4 =

Practice: Multiplication Table 13

Activity 1

Date: ________ Start: ________ Finish: ________ Score: ________

1) 13 X 3 =	2) 13 X 4 =	3) 13 X 6 =
4) 13 X 5 =	5) 13 X 8 =	6) 13 X 10 =
7) 13 X 7 =	8) 13 X 9 =	9) 13 X 2 =
10) 13 X 9 =	11) 13 X 5 =	12) 13 X 2 =
13) 13 X 7 =	14) 13 X 10 =	15) 13 X 3 =
16) 13 X 8 =	17) 13 X 4 =	18) 13 X 6 =
19) 13 X 3 =	20) 13 X 5 =	21) 13 X 4 =

Activity 2

Date: ________ Start: ________ Finish: ________ Score: ________

22) 13 X 7 =	23) 13 X 2 =	24) 13 X 6 =
25) 13 X 8 =	26) 13 X 9 =	27) 13 X 10 =
28) 13 X 4 =	29) 13 X 9 =	30) 13 X 5 =
31) 13 X 6 =	32) 13 X 2 =	33) 13 X 10 =
34) 13 X 3 =	35) 13 X 7 =	36) 13 X 8 =
37) 13 X 2 =	38) 13 X 4 =	39) 13 X 8 =
40) 13 X 9 =	41) 13 X 5 =	42) 13 X 6 =

Activity 1

Date: ______ Start: ______ Finish: ______ Score: ______

1) 14 X 2 =	2) 14 X 5 =	3) 14 X 6 =
4) 14 X 10 =	5) 14 X 3 =	6) 14 X 4 =
7) 14 X 9 =	8) 14 X 7 =	9) 14 X 8 =
10) 14 X 10 =	11) 14 X 2 =	12) 14 X 5 =
13) 14 X 3 =	14) 14 X 6 =	15) 14 X 9 =
16) 14 X 4 =	17) 14 X 8 =	18) 14 X 7 =
19) 14 X 4 =	20) 14 X 6 =	21) 14 X 2 =

Activity 2

Date: ______ Start: ______ Finish: ______ Score: ______

22) 14 X 8 =	23) 14 X 9 =	24) 14 X 5 =
25) 14 X 7 =	26) 14 X 10 =	27) 14 X 3 =
28) 14 X 5 =	29) 14 X 7 =	30) 14 X 8 =
31) 14 X 10 =	32) 14 X 4 =	33) 14 X 9 =
34) 14 X 6 =	35) 14 X 3 =	36) 14 X 2 =
37) 14 X 5 =	38) 14 X 6 =	39) 14 X 10 =
40) 14 X 8 =	41) 14 X 9 =	42) 14 X 2 =

Practice: Multiplication Table 14

Activity 1

Date: ________ Start: ________ Finish: ________ Score: ________

1) 14 X 3 =	2) 14 X 10 =	3) 14 X 6 =
4) 14 X 4 =	5) 14 X 5 =	6) 14 X 8 =
7) 14 X 7 =	8) 14 X 2 =	9) 14 X 9 =
10) 14 X 3 =	11) 14 X 8 =	12) 14 X 4 =
13) 14 X 9 =	14) 14 X 2 =	15) 14 X 5 =
16) 14 X 7 =	17) 14 X 10 =	18) 14 X 6 =
19) 14 X 5 =	20) 14 X 8 =	21) 14 X 2 =

Activity 2

Date: ________ Start: ________ Finish: ________ Score: ________

22) 14 X 4 =	23) 14 X 10 =	24) 14 X 9 =
25) 14 X 3 =	26) 14 X 6 =	27) 14 X 7 =
28) 14 X 8 =	29) 14 X 3 =	30) 14 X 9 =
31) 14 X 2 =	32) 14 X 10 =	33) 14 X 4 =
34) 14 X 7 =	35) 14 X 5 =	36) 14 X 6 =
37) 14 X 3 =	38) 14 X 8 =	39) 14 X 6 =
40) 14 X 4 =	41) 14 X 10 =	42) 14 X 2 =

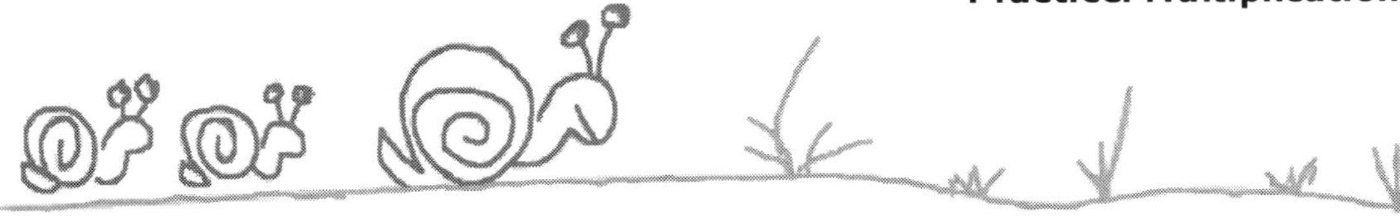

Activity 1

Date: ______ Start: ______ Finish: ______ Score: ______

1 15 X 5 =	2 15 X 7 =	3 15 X 6 =
4 15 X 8 =	5 15 X 9 =	6 15 X 4 =
7 15 X 2 =	8 15 X 10 =	9 15 X 3 =
10 15 X 10 =	11 15 X 5 =	12 15 X 7 =
13 15 X 8 =	14 15 X 3 =	15 15 X 6 =
16 15 X 2 =	17 15 X 4 =	18 15 X 9 =
19 15 X 7 =	20 15 X 5 =	21 15 X 10 =

Activity 2

Date: ______ Start: ______ Finish: ______ Score: ______

22 15 X 8 =	23 15 X 9 =	24 15 X 4 =
25 15 X 2 =	26 15 X 3 =	27 15 X 6 =
28 15 X 4 =	29 15 X 8 =	30 15 X 9 =
31 15 X 6 =	32 15 X 5 =	33 15 X 2 =
34 15 X 10 =	35 15 X 7 =	36 15 X 3 =
37 15 X 5 =	38 15 X 4 =	39 15 X 10 =
40 15 X 9 =	41 15 X 6 =	42 15 X 3 =

Practice: Multiplication Table 15

Activity 1

Date: ________ Start: ________ Finish: ________ Score: ________

1 15 X 4 =	2 15 X 9 =	3 15 X 7 =
4 15 X 5 =	5 15 X 6 =	6 15 X 10 =
7 15 X 3 =	8 15 X 2 =	9 15 X 8 =
10 15 X 4 =	11 15 X 3 =	12 15 X 10 =
13 15 X 5 =	14 15 X 8 =	15 15 X 2 =
16 15 X 9 =	17 15 X 7 =	18 15 X 6 =
19 15 X 4 =	20 15 X 6 =	21 15 X 3 =

Activity 2

Date: ________ Start: ________ Finish: ________ Score: ________

22 15 X 2 =	23 15 X 8 =	24 15 X 5 =
25 15 X 10 =	26 15 X 9 =	27 15 X 7 =
28 15 X 4 =	29 15 X 3 =	30 15 X 7 =
31 15 X 2 =	32 15 X 8 =	33 15 X 5 =
34 15 X 10 =	35 15 X 9 =	36 15 X 6 =
37 15 X 10 =	38 15 X 5 =	39 15 X 2 =
40 15 X 9 =	41 15 X 3 =	42 15 X 7 =

Activity 1

Date: ______ Start: ______ Finish: ______ Score: ______

1	15 X 10 =	2	14 X 2 =	3	12 X 6 =
4	11 X 3 =	5	13 X 8 =	6	12 X 9 =
7	15 X 5 =	8	11 X 7 =	9	13 X 4 =
10	14 X 7 =	11	13 X 6 =	12	14 X 8 =
13	15 X 5 =	14	11 X 2 =	15	12 X 9 =
16	14 X 3 =	17	12 X 4 =	18	13 X 10 =
19	15 X 7 =	20	11 X 10 =	21	13 X 3 =

Activity 2

Date: ______ Start: ______ Finish: ______ Score: ______

22	12 X 4 =	23	11 X 2 =	24	15 X 8 =
25	14 X 6 =	26	15 X 5 =	27	12 X 9 =
28	14 X 3 =	29	11 X 2 =	30	13 X 10 =
31	14 X 7 =	32	11 X 5 =	33	15 X 8 =
34	12 X 6 =	35	13 X 9 =	36	14 X 4 =
37	11 X 3 =	38	12 X 5 =	39	15 X 8 =
40	13 X 10 =	41	12 X 6 =	42	14 X 7 =

Activity 1

Date: ______ Start: ______ Finish: ______ Score: ______

1	12 X 5 =	2	13 X 2 =	3	14 X 10 =
4	15 X 6 =	5	11 X 9 =	6	13 X 3 =
7	14 X 4 =	8	15 X 8 =	9	11 X 7 =
10	12 X 5 =	11	15 X 7 =	12	14 X 4 =
13	11 X 10 =	14	12 X 8 =	15	13 X 2 =
16	11 X 6 =	17	15 X 3 =	18	14 X 9 =
19	13 X 4 =	20	12 X 10 =	21	11 X 9 =

Activity 2

Date: ______ Start: ______ Finish: ______ Score: ______

22	13 X 2 =	23	12 X 5 =	24	14 X 7 =
25	15 X 6 =	26	12 X 3 =	27	15 X 8 =
28	14 X 7 =	29	11 X 6 =	30	13 X 9 =
31	15 X 10 =	32	13 X 5 =	33	14 X 4 =
34	12 X 8 =	35	11 X 3 =	36	15 X 2 =
37	13 X 10 =	38	14 X 9 =	39	12 X 7 =
40	11 X 4 =	41	14 X 2 =	42	11 X 5 =

Activity 1

Date: ______ Start: ______ Finish: ______ Score: ______

1) 11 X 4 =	2) 12 X 8 =	3) 15 X 9 =
4) 13 X 2 =	5) 14 X 10 =	6) 13 X 5 =
7) 12 X 7 =	8) 15 X 6 =	9) 11 X 3 =
10) 14 X 5 =	11) 15 X 10 =	12) 13 X 8 =
13) 11 X 3 =	14) 14 X 9 =	15) 12 X 6 =
16) 14 X 7 =	17) 13 X 2 =	18) 11 X 4 =
19) 15 X 6 =	20) 12 X 8 =	21) 13 X 2 =

Activity 2

Date: ______ Start: ______ Finish: ______ Score: ______

22) 12 X 9 =	23) 11 X 7 =	24) 14 X 5 =
25) 15 X 10 =	26) 12 X 3 =	27) 11 X 4 =
28) 14 X 9 =	29) 13 X 2 =	30) 15 X 7 =
31) 11 X 10 =	32) 13 X 8 =	33) 14 X 6 =
34) 15 X 4 =	35) 12 X 3 =	36) 15 X 5 =
37) 12 X 6 =	38) 13 X 9 =	39) 14 X 3 =
40) 11 X 7 =	41) 14 X 4 =	42) 13 X 2 =

Activity 1

Date: ________ Start: ________ Finish: ________ Score: ________

No.	Problem	No.	Problem	No.	Problem
1	12 X 8 =	2	11 X 9 =	3	14 X 10 =
4	15 X 6 =	5	13 X 7 =	6	11 X 2 =
7	12 X 4 =	8	15 X 3 =	9	13 X 5 =
10	14 X 10 =	11	13 X 5 =	12	12 X 6 =
13	15 X 2 =	14	11 X 8 =	15	14 X 4 =
16	15 X 3 =	17	13 X 9 =	18	12 X 7 =
19	11 X 4 =	20	14 X 6 =	21	13 X 10 =

Activity 2

Date: ________ Start: ________ Finish: ________ Score: ________

No.	Problem	No.	Problem	No.	Problem
22	14 X 5 =	23	11 X 2 =	24	12 X 9 =
25	15 X 7 =	26	11 X 8 =	27	12 X 3 =
28	14 X 7 =	29	13 X 6 =	30	15 X 4 =
31	13 X 5 =	32	11 X 2 =	33	14 X 9 =
34	12 X 3 =	35	15 X 10 =	36	14 X 8 =
37	12 X 7 =	38	15 X 2 =	39	11 X 4 =
40	13 X 10 =	41	12 X 3 =	42	14 X 5 =

Activity 1

Date: ______ Start: ______ Finish: ______ Score: ______

1 16 X 7 =	2 16 X 2 =	3 16 X 5 =
4 16 X 8 =	5 16 X 10 =	6 16 X 9 =
7 16 X 6 =	8 16 X 3 =	9 16 X 4 =
10 16 X 6 =	11 16 X 4 =	12 16 X 5 =
13 16 X 7 =	14 16 X 3 =	15 16 X 2 =
16 16 X 9 =	17 16 X 10 =	18 16 X 8 =
19 16 X 4 =	20 16 X 2 =	21 16 X 6 =

Activity 2

Date: ______ Start: ______ Finish: ______ Score: ______

22 16 X 7 =	23 16 X 10 =	24 16 X 3 =
25 16 X 8 =	26 16 X 9 =	27 16 X 5 =
28 16 X 3 =	29 16 X 6 =	30 16 X 8 =
31 16 X 9 =	32 16 X 7 =	33 16 X 10 =
34 16 X 2 =	35 16 X 5 =	36 16 X 4 =
37 16 X 9 =	38 16 X 7 =	39 16 X 2 =
40 16 X 6 =	41 16 X 10 =	42 16 X 4 =

Activity 1

Date: ______ Start: ______ Finish: ______ Score: ______

1. 16 X 7 =	2. 16 X 5 =	3. 16 X 6 =
4. 16 X 2 =	5. 16 X 3 =	6. 16 X 10 =
7. 16 X 8 =	8. 16 X 4 =	9. 16 X 9 =
10. 16 X 10 =	11. 16 X 9 =	12. 16 X 5 =
13. 16 X 3 =	14. 16 X 6 =	15. 16 X 4 =
16. 16 X 7 =	17. 16 X 2 =	18. 16 X 8 =
19. 16 X 4 =	20. 16 X 2 =	21. 16 X 5 =

Activity 2

Date: ______ Start: ______ Finish: ______ Score: ______

22. 16 X 7 =	23. 16 X 8 =	24. 16 X 10 =
25. 16 X 9 =	26. 16 X 6 =	27. 16 X 3 =
28. 16 X 4 =	29. 16 X 10 =	30. 16 X 7 =
31. 16 X 8 =	32. 16 X 5 =	33. 16 X 9 =
34. 16 X 2 =	35. 16 X 6 =	36. 16 X 3 =
37. 16 X 10 =	38. 16 X 3 =	39. 16 X 4 =
40. 16 X 5 =	41. 16 X 8 =	42. 16 X 9 =

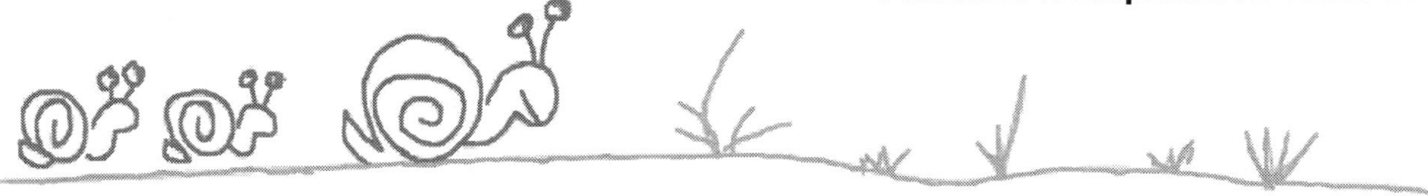

Activity 1

Date: ______ Start: ______ Finish: ______ Score: ______

1 17 X 8 =	2 17 X 3 =	3 17 X 9 =
4 17 X 7 =	5 17 X 10 =	6 17 X 6 =
7 17 X 4 =	8 17 X 5 =	9 17 X 2 =
10 17 X 7 =	11 17 X 6 =	12 17 X 10 =
13 17 X 8 =	14 17 X 9 =	15 17 X 4 =
16 17 X 5 =	17 17 X 3 =	18 17 X 2 =
19 17 X 7 =	20 17 X 4 =	21 17 X 10 =

Activity 2

Date: ______ Start: ______ Finish: ______ Score: ______

22 17 X 2 =	23 17 X 6 =	24 17 X 5 =
25 17 X 8 =	26 17 X 9 =	27 17 X 3 =
28 17 X 2 =	29 17 X 5 =	30 17 X 6 =
31 17 X 9 =	32 17 X 4 =	33 17 X 10 =
34 17 X 8 =	35 17 X 7 =	36 17 X 3 =
37 17 X 9 =	38 17 X 5 =	39 17 X 8 =
40 17 X 3 =	41 17 X 10 =	42 17 X 4 =

Practice: Multiplication Table 17

Activity 1

Date: ________ Start: ________ Finish: ________ Score: ________

1. 17 X 10 =	2. 17 X 8 =	3. 17 X 9 =
4. 17 X 6 =	5. 17 X 2 =	6. 17 X 4 =
7. 17 X 3 =	8. 17 X 7 =	9. 17 X 5 =
10. 17 X 4 =	11. 17 X 7 =	12. 17 X 5 =
13. 17 X 9 =	14. 17 X 8 =	15. 17 X 2 =
16. 17 X 10 =	17. 17 X 6 =	18. 17 X 3 =
19. 17 X 8 =	20. 17 X 6 =	21. 17 X 4 =

Activity 2

Date: ________ Start: ________ Finish: ________ Score: ________

22. 17 X 5 =	23. 17 X 2 =	24. 17 X 9 =
25. 17 X 3 =	26. 17 X 10 =	27. 17 X 7 =
28. 17 X 6 =	29. 17 X 9 =	30. 17 X 3 =
31. 17 X 4 =	32. 17 X 10 =	33. 17 X 2 =
34. 17 X 8 =	35. 17 X 7 =	36. 17 X 5 =
37. 17 X 3 =	38. 17 X 8 =	39. 17 X 10 =
40. 17 X 4 =	41. 17 X 6 =	42. 17 X 2 =

Activity 1

Date: ________ Start: ________ Finish: ________ Score: ________

1. 18 X 10 = ☐	2. 18 X 5 = ☐	3. 18 X 6 = ☐
4. 18 X 2 = ☐	5. 18 X 7 = ☐	6. 18 X 9 = ☐
7. 18 X 4 = ☐	8. 18 X 3 = ☐	9. 18 X 8 = ☐
10. 18 X 6 = ☐	11. 18 X 2 = ☐	12. 18 X 4 = ☐
13. 18 X 5 = ☐	14. 18 X 3 = ☐	15. 18 X 10 = ☐
16. 18 X 9 = ☐	17. 18 X 8 = ☐	18. 18 X 7 = ☐
19. 18 X 2 = ☐	20. 18 X 5 = ☐	21. 18 X 9 = ☐

Activity 2

Date: ________ Start: ________ Finish: ________ Score: ________

22. 18 X 4 = ☐	23. 18 X 10 = ☐	24. 18 X 3 = ☐
25. 18 X 8 = ☐	26. 18 X 7 = ☐	27. 18 X 6 = ☐
28. 18 X 5 = ☐	29. 18 X 3 = ☐	30. 18 X 7 = ☐
31. 18 X 8 = ☐	32. 18 X 2 = ☐	33. 18 X 10 = ☐
34. 18 X 6 = ☐	35. 18 X 9 = ☐	36. 18 X 4 = ☐
37. 18 X 3 = ☐	38. 18 X 9 = ☐	39. 18 X 5 = ☐
40. 18 X 8 = ☐	41. 18 X 4 = ☐	42. 18 X 7 = ☐

Practice: Multiplication Table 18

Activity 1

Date: ________ Start: ________ Finish: ________ Score: ________

1. 18 X 7 =	2. 18 X 3 =	3. 18 X 2 =
4. 18 X 10 =	5. 18 X 9 =	6. 18 X 6 =
7. 18 X 5 =	8. 18 X 4 =	9. 18 X 8 =
10. 18 X 5 =	11. 18 X 3 =	12. 18 X 10 =
13. 18 X 4 =	14. 18 X 7 =	15. 18 X 8 =
16. 18 X 2 =	17. 18 X 6 =	18. 18 X 9 =
19. 18 X 6 =	20. 18 X 5 =	21. 18 X 8 =

Activity 2

Date: ________ Start: ________ Finish: ________ Score: ________

22. 18 X 2 =	23. 18 X 4 =	24. 18 X 9 =
25. 18 X 10 =	26. 18 X 3 =	27. 18 X 7 =
28. 18 X 9 =	29. 18 X 3 =	30. 18 X 8 =
31. 18 X 2 =	32. 18 X 10 =	33. 18 X 5 =
34. 18 X 7 =	35. 18 X 6 =	36. 18 X 4 =
37. 18 X 9 =	38. 18 X 2 =	39. 18 X 4 =
40. 18 X 6 =	41. 18 X 3 =	42. 18 X 8 =

Activity 1

Date: ______ Start: ______ Finish: ______ Score: ______

1) 19 X 10 =	2) 19 X 8 =	3) 19 X 3 =
4) 19 X 4 =	5) 19 X 7 =	6) 19 X 2 =
7) 19 X 6 =	8) 19 X 9 =	9) 19 X 5 =
10) 19 X 6 =	11) 19 X 5 =	12) 19 X 4 =
13) 19 X 9 =	14) 19 X 8 =	15) 19 X 2 =
16) 19 X 10 =	17) 19 X 7 =	18) 19 X 3 =
19) 19 X 8 =	20) 19 X 9 =	21) 19 X 4 =

Activity 2

Date: ______ Start: ______ Finish: ______ Score: ______

22) 19 X 10 =	23) 19 X 5 =	24) 19 X 3 =
25) 19 X 2 =	26) 19 X 7 =	27) 19 X 6 =
28) 19 X 5 =	29) 19 X 9 =	30) 19 X 2 =
31) 19 X 7 =	32) 19 X 8 =	33) 19 X 3 =
34) 19 X 10 =	35) 19 X 4 =	36) 19 X 6 =
37) 19 X 4 =	38) 19 X 9 =	39) 19 X 10 =
40) 19 X 2 =	41) 19 X 6 =	42) 19 X 7 =

Practice: Multiplication Table 19

Activity 1

Date: ______ Start: ______ Finish: ______ Score: ______

1. 19 X 6 =	2. 19 X 2 =	3. 19 X 5 =
4. 19 X 4 =	5. 19 X 3 =	6. 19 X 8 =
7. 19 X 9 =	8. 19 X 7 =	9. 19 X 10 =
10. 19 X 4 =	11. 19 X 10 =	12. 19 X 8 =
13. 19 X 6 =	14. 19 X 2 =	15. 19 X 7 =
16. 19 X 3 =	17. 19 X 9 =	18. 19 X 5 =
19. 19 X 7 =	20. 19 X 10 =	21. 19 X 3 =

Activity 2

Date: ______ Start: ______ Finish: ______ Score: ______

22. 19 X 4 =	23. 19 X 8 =	24. 19 X 5 =
25. 19 X 9 =	26. 19 X 2 =	27. 19 X 6 =
28. 19 X 4 =	29. 19 X 8 =	30. 19 X 6 =
31. 19 X 2 =	32. 19 X 5 =	33. 19 X 3 =
34. 19 X 7 =	35. 19 X 10 =	36. 19 X 9 =
37. 19 X 6 =	38. 19 X 8 =	39. 19 X 9 =
40. 19 X 5 =	41. 19 X 7 =	42. 19 X 4 =

Activity 1

Date: ______ Start: ______ Finish: ______ Score: ______

1 20 X 5 =	2 20 X 7 =	3 20 X 10 =
4 20 X 2 =	5 20 X 3 =	6 20 X 4 =
7 20 X 9 =	8 20 X 8 =	9 20 X 6 =
10 20 X 10 =	11 20 X 8 =	12 20 X 4 =
13 20 X 9 =	14 20 X 7 =	15 20 X 6 =
16 20 X 2 =	17 20 X 5 =	18 20 X 3 =
19 20 X 9 =	20 20 X 10 =	21 20 X 3 =

Activity 2

Date: ______ Start: ______ Finish: ______ Score: ______

22 20 X 6 =	23 20 X 8 =	24 20 X 7 =
25 20 X 2 =	26 20 X 5 =	27 20 X 4 =
28 20 X 10 =	29 20 X 4 =	30 20 X 2 =
31 20 X 3 =	32 20 X 9 =	33 20 X 6 =
34 20 X 8 =	35 20 X 7 =	36 20 X 5 =
37 20 X 8 =	38 20 X 3 =	39 20 X 5 =
40 20 X 6 =	41 20 X 10 =	42 20 X 2 =

Activity 1

Date: ________ Start: ________ Finish: ________ Score: ________

1) 20 X 4 =	2) 20 X 3 =	3) 20 X 10 =
4) 20 X 6 =	5) 20 X 2 =	6) 20 X 7 =
7) 20 X 9 =	8) 20 X 5 =	9) 20 X 8 =
10) 20 X 6 =	11) 20 X 4 =	12) 20 X 7 =
13) 20 X 5 =	14) 20 X 8 =	15) 20 X 3 =
16) 20 X 9 =	17) 20 X 10 =	18) 20 X 2 =
19) 20 X 10 =	20) 20 X 4 =	21) 20 X 9 =

Activity 2

Date: ________ Start: ________ Finish: ________ Score: ________

22) 20 X 6 =	23) 20 X 3 =	24) 20 X 5 =
25) 20 X 2 =	26) 20 X 7 =	27) 20 X 8 =
28) 20 X 2 =	29) 20 X 7 =	30) 20 X 8 =
31) 20 X 5 =	32) 20 X 4 =	33) 20 X 6 =
34) 20 X 3 =	35) 20 X 10 =	36) 20 X 9 =
37) 20 X 2 =	38) 20 X 9 =	39) 20 X 3 =
40) 20 X 4 =	41) 20 X 8 =	42) 20 X 6 =

Activity 1

Date: ______ Start: ______ Finish: ______ Score: ______

1	18 X 10 =	2	16 X 8 =	3	19 X 7 =
4	20 X 2 =	5	17 X 6 =	6	19 X 4 =
7	18 X 5 =	8	16 X 3 =	9	17 X 9 =
10	20 X 6 =	11	19 X 8 =	12	20 X 3 =
13	17 X 7 =	14	18 X 9 =	15	16 X 10 =
16	17 X 5 =	17	18 X 4 =	18	19 X 2 =
19	16 X 10 =	20	20 X 6 =	21	16 X 8 =

Activity 2

Date: ______ Start: ______ Finish: ______ Score: ______

22	18 X 4 =	23	19 X 7 =	24	17 X 2 =
25	20 X 3 =	26	16 X 9 =	27	17 X 5 =
28	18 X 7 =	29	20 X 8 =	30	19 X 4 =
31	20 X 5 =	32	17 X 2 =	33	19 X 9 =
34	18 X 6 =	35	16 X 3 =	36	19 X 10 =
37	20 X 2 =	38	18 X 5 =	39	17 X 10 =
40	16 X 6 =	41	18 X 7 =	42	19 X 3 =

Review: Multiplication Table 16 to 20 Mixed

Activity 1

Date: ______ Start: ______ Finish: ______ Score: ______

1. 19 X 3 =
2. 18 X 8 =
3. 20 X 10 =
4. 16 X 5 =
5. 17 X 9 =
6. 18 X 2 =
7. 19 X 4 =
8. 20 X 7 =
9. 16 X 6 =
10. 17 X 9 =
11. 19 X 8 =
12. 16 X 6 =
13. 20 X 10 =
14. 18 X 4 =
15. 17 X 2 =
16. 16 X 5 =
17. 19 X 3 =
18. 18 X 7 =
19. 17 X 5 =
20. 20 X 3 =
21. 16 X 2 =

Activity 2

Date: ______ Start: ______ Finish: ______ Score: ______

22. 19 X 9 =
23. 18 X 6 =
24. 20 X 4 =
25. 17 X 10 =
26. 20 X 7 =
27. 17 X 8 =
28. 16 X 9 =
29. 18 X 5 =
30. 19 X 7 =
31. 17 X 4 =
32. 16 X 10 =
33. 18 X 2 =
34. 19 X 8 =
35. 20 X 6 =
36. 19 X 3 =
37. 17 X 9 =
38. 16 X 8 =
39. 18 X 6 =
40. 20 X 4 =
41. 16 X 5 =
42. 19 X 7 =

Activity 1

Date: ______ Start: ______ Finish: ______ Score: ______

1. 20 X 2 =	2. 16 X 3 =	3. 17 X 6 =
4. 19 X 4 =	5. 18 X 10 =	6. 20 X 5 =
7. 18 X 9 =	8. 17 X 8 =	9. 16 X 7 =
10. 19 X 6 =	11. 16 X 10 =	12. 18 X 7 =
13. 19 X 4 =	14. 17 X 2 =	15. 20 X 3 =
16. 17 X 9 =	17. 16 X 8 =	18. 18 X 5 =
19. 19 X 7 =	20. 20 X 8 =	21. 17 X 5 =

Activity 2

Date: ______ Start: ______ Finish: ______ Score: ______

22. 20 X 6 =	23. 18 X 9 =	24. 19 X 3 =
25. 16 X 4 =	26. 19 X 10 =	27. 20 X 2 =
28. 18 X 8 =	29. 16 X 10 =	30. 17 X 9 =
31. 20 X 5 =	32. 16 X 6 =	33. 17 X 7 =
34. 18 X 4 =	35. 19 X 2 =	36. 20 X 3 =
37. 17 X 6 =	38. 18 X 7 =	39. 16 X 10 =
40. 19 X 9 =	41. 17 X 4 =	42. 16 X 3 =

Review: Multiplication Table 16 to 20 Mixed

Activity 1

Date: ________ Start: ________ Finish: ________ Score: ________

1. 18 X 7 = ☐	2. 17 X 8 = ☐	3. 16 X 6 = ☐
4. 20 X 3 = ☐	5. 19 X 10 = ☐	6. 16 X 2 = ☐
7. 20 X 9 = ☐	8. 17 X 5 = ☐	9. 18 X 4 = ☐
10. 19 X 9 = ☐	11. 17 X 10 = ☐	12. 19 X 4 = ☐
13. 20 X 5 = ☐	14. 16 X 2 = ☐	15. 18 X 7 = ☐
16. 17 X 8 = ☐	17. 20 X 3 = ☐	18. 18 X 6 = ☐
19. 19 X 5 = ☐	20. 16 X 2 = ☐	21. 20 X 6 = ☐

Activity 2

Date: ________ Start: ________ Finish: ________ Score: ________

22. 16 X 4 = ☐	23. 18 X 8 = ☐	24. 19 X 7 = ☐
25. 17 X 3 = ☐	26. 16 X 10 = ☐	27. 20 X 9 = ☐
28. 17 X 6 = ☐	29. 18 X 4 = ☐	30. 19 X 2 = ☐
31. 17 X 3 = ☐	32. 16 X 10 = ☐	33. 18 X 7 = ☐
34. 19 X 5 = ☐	35. 20 X 8 = ☐	36. 18 X 9 = ☐
37. 16 X 10 = ☐	38. 19 X 5 = ☐	39. 17 X 2 = ☐
40. 20 X 9 = ☐	41. 17 X 6 = ☐	42. 18 X 3 = ☐

Activity 1

Date: Start: Finish: Score:

1	2 X 8 =	2	19 X 3 =	3	20 X 4 =
4	3 X 5 =	5	16 X 6 =	6	8 X 10 =
7	7 X 7 =	8	17 X 2 =	9	18 X 9 =
10	10 X 10 =	11	13 X 3 =	12	14 X 5 =
13	4 X 4 =	14	6 X 7 =	15	11 X 9 =
16	15 X 8 =	17	5 X 6 =	18	12 X 2 =
19	9 X 7 =	20	16 X 8 =	21	9 X 10 =

Activity 2

Date: Start: Finish: Score:

22	20 X 5 =	23	10 X 4 =	24	5 X 3 =
25	18 X 2 =	26	13 X 6 =	27	11 X 9 =
28	14 X 10 =	29	12 X 9 =	30	7 X 6 =
31	15 X 7 =	32	19 X 4 =	33	4 X 3 =
34	6 X 2 =	35	17 X 5 =	36	3 X 8 =
37	2 X 5 =	38	8 X 9 =	39	3 X 7 =
40	8 X 8 =	41	4 X 6 =	42	10 X 10 =

Activity 1

Date: ______ Start: ______ Finish: ______ Score: ______

1	7 X 7 =	2	12 X 9 =	3	4 X 8 =
4	15 X 10 =	5	2 X 4 =	6	18 X 5 =
7	9 X 6 =	8	10 X 2 =	9	13 X 3 =
10	14 X 10 =	11	8 X 4 =	12	5 X 6 =
13	6 X 9 =	14	17 X 2 =	15	19 X 7 =
16	16 X 8 =	17	3 X 5 =	18	20 X 3 =
19	11 X 6 =	20	9 X 9 =	21	13 X 10 =

Activity 2

Date: ______ Start: ______ Finish: ______ Score: ______

22	3 X 2 =	23	8 X 4 =	24	4 X 3 =
25	17 X 5 =	26	10 X 7 =	27	11 X 8 =
28	19 X 6 =	29	5 X 5 =	30	18 X 10 =
31	6 X 9 =	32	14 X 3 =	33	12 X 4 =
34	7 X 2 =	35	2 X 7 =	36	15 X 8 =
37	20 X 3 =	38	16 X 8 =	39	11 X 2 =
40	9 X 9 =	41	2 X 10 =	42	7 X 6 =

Activity 1

Date: ______ Start: ______ Finish: ______ Score: ______

1	14 X 6 =	2	8 X 4 =	3	17 X 9 =
4	6 X 2 =	5	16 X 10 =	6	11 X 8 =
7	10 X 5 =	8	9 X 3 =	9	5 X 7 =
10	2 X 3 =	11	12 X 5 =	12	4 X 4 =
13	18 X 6 =	14	20 X 2 =	15	19 X 10 =
16	13 X 7 =	17	3 X 9 =	18	15 X 8 =
19	7 X 2 =	20	11 X 5 =	21	19 X 7 =

Activity 2

Date: ______ Start: ______ Finish: ______ Score: ______

22	12 X 10 =	23	18 X 9 =	24	17 X 3 =
25	9 X 8 =	26	6 X 4 =	27	13 X 6 =
28	4 X 3 =	29	3 X 7 =	30	14 X 9 =
31	8 X 5 =	32	10 X 2 =	33	7 X 4 =
34	5 X 8 =	35	2 X 10 =	36	15 X 6 =
37	20 X 4 =	38	16 X 3 =	39	7 X 5 =
40	12 X 7 =	41	14 X 2 =	42	9 X 9 =

Activity 1

Date: ______ Start: ______ Finish: ______ Score: ______

1	11 X 3 =	2	16 X 4 =	3	7 X 6 =
4	17 X 5 =	5	4 X 9 =	6	2 X 10 =
7	13 X 2 =	8	20 X 7 =	9	19 X 8 =
10	6 X 9 =	11	18 X 3 =	12	10 X 6 =
13	3 X 8 =	14	8 X 2 =	15	5 X 10 =
16	15 X 7 =	17	14 X 4 =	18	12 X 5 =
19	9 X 3 =	20	20 X 8 =	21	14 X 10 =

Activity 2

Date: ______ Start: ______ Finish: ______ Score: ______

22	3 X 7 =	23	5 X 9 =	24	8 X 2 =
25	4 X 5 =	26	16 X 6 =	27	18 X 4 =
28	19 X 9 =	29	13 X 6 =	30	17 X 4 =
31	9 X 7 =	32	6 X 5 =	33	10 X 3 =
34	15 X 8 =	35	11 X 2 =	36	7 X 10 =
37	12 X 2 =	38	2 X 8 =	39	6 X 10 =
40	19 X 5 =	41	3 X 4 =	42	14 X 7 =

Page 5

1. 2, 3, 2x3=6
2. 4, 2, 4x2=8
3. 3, 7, 3x7=21

Page 6

1. 3x3=9
2. 4x3=12
3. 3x1=3
4. 2x3=6
5. 5x6=30

Page 9

1. 6, 12, 14
2. 18, 27, 30, 36
3. 8, 12, 24, 28, 32
4. 5, 15, 20, 25, 30, 35
5. 24, 30, 36, 46, 52, 58

Page 11

1. 3x4 = 4+4+4 = 12
2. 5x2 = 2+2+2+2+2 = 10
3. 2x6 = 6+6 = 12
4. 4x4 = 4+4+4+4 = 16
5. 3x10 = 10+10+10 = 30
6. 5x8 = 8+8+8+8+8 = 40

Page 18

1. 0
2. 10
3. 0
4. 2
5. 0
6. 6
7. 0
8. 4
9. 0
10. 8
11. 0
12. 2
13. 0
14. 3
15. 0
16. 5
17. 0
18. 9
19. 0
20. 8
21. 0
22. 7
23. 0
24. 3
25. 0
26. 5
27. 0
28. 2
29. 0
30. 10
31. 0
32. 8
33. 0
34. 7
35. 0
36. 5
37. 0
38. 6
39. 0
40. 8
41. 0
42. 3

Page 19

1. 4
2. 8
3. 6
4. 18
5. 16
6. 12
7. 14
8. 20
9. 10
10. 8
11. 18
12. 16
13. 6
14. 14
15. 4
16. 20
17. 10
18. 12
19. 6
20. 14
21. 16
22. 20
23. 12
24. 10
25. 4
26. 8
27. 18
28. 10
29. 18
30. 8
31. 6
32. 12
33. 4
34. 20
35. 14
36. 16
37. 10
38. 16
39. 6
40. 4
41. 12
42. 18

Page 20

1. 12
2. 20
3. 6
4. 8
5. 16
6. 18
7. 14
8. 10
9. 4
10. 6
11. 4
12. 8
13. 16
14. 20
15. 18
16. 12
17. 10
18. 14
19. 18
20. 14
21. 16
22. 8
23. 4
24. 20
25. 12
26. 6
27. 10
28. 4
29. 16
30. 14
31. 18
32. 20
33. 10
34. 12
35. 8
36. 6
37. 18
38. 12
39. 4
40. 8
41. 16
42. 10

Page 21

1. 6
2. 14
3. 20
4. 10
5. 8
6. 16
7. 4
8. 18
9. 12
10. 4
11. 16
12. 18
13. 12
14. 6
15. 20
16. 8
17. 14
18. 10
19. 16
20. 10
21. 18
22. 20
23. 4
24. 12
25. 14
26. 8
27. 6
28. 10
29. 8
30. 12
31. 14
32. 20
33. 6
34. 16
35. 4
36. 18
37. 20
38. 4
39. 18
40. 14
41. 16
42. 6

Page 22

1. 27
2. 18
3. 9
4. 30
5. 24
6. 21
7. 15
8. 12
9. 6
10. 18
11. 21
12. 12
13. 9
14. 30
15. 6
16. 15
17. 27
18. 24
19. 27
20. 24
21. 30
22. 15
23. 18
24. 6
25. 21
26. 12
27. 9
28. 15
29. 30
30. 12
31. 24
32. 27
33. 21
34. 18
35. 9
36. 6
37. 9
38. 24
39. 27
40. 18
41. 12
42. 30

Answer Key

Page 23

1. 27
2. 9
3. 15
4. 12
5. 30
6. 6
7. 21
8. 18
9. 24
10. 15
11. 6
12. 30
13. 21
14. 12
15. 27
16. 18
17. 9
18. 24
19. 15
20. 21
21. 12
22. 9
23. 6
24. 30
25. 24
26. 18
27. 27
28. 30
29. 21
30. 24
31. 6
32. 27
33. 18
34. 12
35. 15
36. 9
37. 18
38. 12
39. 27
40. 9
41. 21
42. 15

Page 24

1. 27
2. 21
3. 15
4. 6
5. 30
6. 24
7. 18
8. 9
9. 12
10. 6
11. 18
12. 15
13. 30
14. 9
15. 24
16. 12
17. 21
18. 27
19. 12
20. 9
21. 15
22. 27
23. 21
24. 24
25. 6
26. 30
27. 18
28. 12
29. 24
30. 15
31. 30
32. 21
33. 9
34. 18
35. 6
36. 27
37. 30
38. 21
39. 18
40. 27
41. 24
42. 6

Page 25

1. 20
2. 16
3. 36
4. 40
5. 32
6. 8
7. 24
8. 28
9. 12
10. 40
11. 12
12. 16
13. 28
14. 36
15. 32
16. 24
17. 20
18. 8
19. 20
20. 24
21. 16
22. 36
23. 40
24. 12
25. 28
26. 8
27. 32
28. 28
29. 8
30. 16
31. 32
32. 20
33. 36
34. 24
35. 12
36. 40
37. 28
38. 12
39. 20
40. 24
41. 16
42. 36

Page 26

1. 8
2. 36
3. 20
4. 40
5. 16
6. 28
7. 24
8. 12
9. 32
10. 12
11. 8
12. 32
13. 40
14. 16
15. 24
16. 20
17. 28
18. 36
19. 8
20. 28
21. 32
22. 40
23. 16
24. 36
25. 24
26. 12
27. 20
28. 32
29. 20
30. 40
31. 24
32. 16
33. 12
34. 36
35. 8
36. 28
37. 12
38. 16
39. 28
40. 24
41. 36
42. 40

Page 27

1. 8
2. 28
3. 16
4. 36
5. 20
6. 12
7. 32
8. 40
9. 24
10. 28
11. 24
12. 36
13. 12
14. 16
15. 8
16. 40
17. 20
18. 32
19. 28
20. 40
21. 24
22. 36
23. 12
24. 8
25. 32
26. 20
27. 16
28. 20
29. 36
30. 28
31. 12
32. 8
33. 16
34. 40
35. 24
36. 32
37. 40
38. 28
39. 8
40. 24
41. 16
42. 36

Page 28

1. 25
2. 40
3. 35
4. 15
5. 20
6. 10
7. 50
8. 45
9. 30
10. 15
11. 35
12. 40
13. 25
14. 45
15. 30
16. 20
17. 50
18. 10
19. 15
20. 50
21. 30
22. 25
23. 40
24. 20
25. 35
26. 45
27. 10
28. 45
29. 50
30. 35
31. 10
32. 40
33. 15
34. 20
35. 30
36. 25
37. 10
38. 30
39. 15
40. 45
41. 35
42. 50

Page 29

1. 15
2. 25
3. 30
4. 40
5. 10
6. 35
7. 45
8. 50
9. 20
10. 35
11. 50
12. 20
13. 15

14. 30
15. 25
16. 40
17. 10
18. 45
19. 25
20. 35
21. 45
22. 10
23. 15
24. 50
25. 30
26. 40
27. 20
28. 25
29. 10
30. 30
31. 35
32. 20
33. 45
34. 50
35. 15
36. 40
37. 50
38. 45
39. 30
40. 35
41. 10
42. 15

Page 30

1. 25
2. 40
3. 10
4. 35
5. 20
6. 45
7. 15
8. 50
9. 30
10. 45
11. 50
12. 35
13. 25
14. 30
15. 10
16. 20
17. 15
18. 40
19. 15
20. 40
21. 35
22. 10
23. 45
24. 20
25. 25
26. 30
27. 50
28. 25
29. 35
30. 10
31. 30
32. 40
33. 50
34. 45
35. 15
36. 20
37. 30
38. 40
39. 20
40. 45
41. 25
42. 35

Page 31

1. 27
2. 20
3. 50
4. 8
5. 15
6. 6
7. 28
8. 16
9. 18
10. 20
11. 10
12. 24
13. 14
14. 12
15. 20
16. 27
17. 32
18. 10
19. 35
20. 27
21. 12
22. 30
23. 10
24. 24
25. 50
26. 16
27. 6
28. 16
29. 45
30. 12
31. 24
32. 14
33. 15
34. 20
35. 10
36. 12
37. 12
38. 6
39. 36
40. 50
41. 20
42. 6

Page 32

1. 15
2. 32
3. 20
4. 6
5. 28
6. 20
7. 27
8. 30
9. 6
10. 15
11. 28
12. 4
13. 24
14. 24
15. 20
16. 20
17. 25
18. 18
19. 40
20. 9
21. 24
22. 16
23. 6
24. 35
25. 36
26. 15
27. 20
28. 4
29. 9
30. 20
31. 18
32. 32
33. 15
34. 14
35. 50
36. 24
37. 50
38. 4
39. 18
40. 12
41. 12
42. 28

Page 33

1. 4
2. 30
3. 32
4. 20
5. 24
6. 27
7. 25
8. 14
9. 9
10. 4
11. 20
12. 40
13. 16
14. 24
15. 35
16. 15
17. 36
18. 6
19. 12
20. 40
21. 36
22. 6
23. 10
24. 21
25. 30
26. 10
27. 40
28. 12
29. 10
30. 10
31. 9
32. 24
33. 27
34. 20
35. 40
36. 28
37. 27
38. 16
39. 20
40. 28
41. 4
42. 12

Page 34

1. 10
2. 6
3. 50
4. 24
5. 0
6. 27
7. 32
8. 35
9. 16
10. 12
11. 20
12. 15
13. 8
14. 27
15. 6
16. 50
17. 14
18. 40
19. 16
20. 15
21. 40
22. 21
23. 18
24. 30
25. 4
26. 15
27. 32
28. 21
29. 6
30. 20
31. 15
32. 24
33. 24
34. 40
35. 10
36. 18
37. 16
38. 30
39. 18
40. 25
41. 16
42. 30

Page 35

1. 15
2. 6
3. 28
4. 20
5. 27
6. 30
7. 32
8. 20
9. 10
10. 32
11. 20
12. 6
13. 15
14. 28
15. 18
16. 12
17. 10
18. 24
19. 27
20. 15
21. 28
22. 24
23. 30
24. 10
25. 12
26. 4
27. 40

28. 20
29. 10
30. 24
31. 15
32. 36
33. 30
34. 40
35. 21
36. 4
37. 30
38. 28
39. 15
40. 6
41. 32
42. 27

Page 36

1. 12
2. 24
3. 54
4. 18
5. 36
6. 42
7. 48
8. 60
9. 30
10. 18
11. 42
12. 54
13. 48
14. 12
15. 60
16. 30
17. 36
18. 24
19. 18
20. 60
21. 30
22. 36
23. 12
24. 48
25. 42
26. 24
27. 54
28. 42
29. 60
30. 48
31. 12
32. 30
33. 36
34. 18
35. 24
36. 54
37. 60
38. 12
39. 18
40. 42
41. 36
42. 30

Page 37

1. 12
2. 24
3. 48
4. 42
5. 60
6. 30
7. 18
8. 54
9. 36
10. 18
11. 42
12. 24
13. 54
14. 36
15. 30
16. 48
17. 12
18. 60
19. 30
20. 18
21. 60
22. 48
23. 42
24. 54
25. 24
26. 36
27. 12
28. 54
29. 30
30. 36
31. 12
32. 42
33. 48
34. 60
35. 24
36. 18
37. 30
38. 42
39. 12
40. 60
41. 24
42. 48

Page 38

1. 12
2. 18
3. 42
4. 30
5. 54
6. 36
7. 24
8. 48
9. 60
10. 36
11. 48
12. 42
13. 60
14. 24
15. 30
16. 18
17. 12
18. 54
19. 48
20. 12
21. 42
22. 60
23. 30
24. 18
25. 24
26. 54
27. 36
28. 30
29. 42
30. 24
31. 36
32. 48
33. 54
34. 12
35. 60
36. 18
37. 12
38. 18
39. 36
40. 42
41. 60
42. 24

Page 39

1. 21
2. 49
3. 35
4. 63
5. 56
6. 42
7. 14
8. 70
9. 28
10. 35
11. 49
12. 70
13. 63
14. 42
15. 14
16. 56
17. 21
18. 28
19. 14
20. 70
21. 42
22. 49
23. 63
24. 21
25. 28
26. 35
27. 56
28. 35
29. 70
30. 14
31. 63
32. 28
33. 21
34. 42
35. 49
36. 56
37. 28
38. 63
39. 49
40. 42
41. 35
42. 21

Page 40

1. 42
2. 56
3. 63
4. 49
5. 21
6. 35
7. 14
8. 28
9. 70
10. 56
11. 21
12. 42
13. 70
14. 28
15. 63
16. 35
17. 14
18. 49
19. 21
20. 42
21. 49
22. 28
23. 56
24. 63
25. 14
26. 35
27. 70
28. 49
29. 70
30. 63
31. 21
32. 14
33. 42
34. 35
35. 56
36. 28
37. 35
38. 21
39. 14
40. 70
41. 56
42. 49

Page 41

1. 70
2. 42
3. 56
4. 35
5. 49
6. 14
7. 28
8. 63
9. 21
10. 63
11. 70
12. 42
13. 49
14. 14
15. 28
16. 56
17. 21
18. 35
19. 70
20. 28
21. 42
22. 21
23. 14
24. 35
25. 63
26. 49
27. 56
28. 21
29. 28
30. 63
31. 56
32. 49
33. 14
34. 42
35. 35
36. 70
37. 28
38. 70
39. 56
40. 35

41. 42
42. 21

Page 42

1. 16
2. 48
3. 40
4. 80
5. 32
6. 64
7. 24
8. 56
9. 72
10. 56
11. 72
12. 48
13. 80
14. 32
15. 24
16. 40
17. 64
18. 16
19. 48
20. 32
21. 72
22. 64
23. 56
24. 80
25. 40
26. 16
27. 24
28. 56
29. 40
30. 72
31. 48
32. 16
33. 24
34. 64
35. 80
36. 32
37. 24
38. 56
39. 48
40. 80
41. 40
42. 64

Page 43

1. 32
2. 56
3. 72
4. 16
5. 64
6. 24
7. 80
8. 40
9. 48
10. 64
11. 48
12. 24
13. 56
14. 80
15. 72
16. 40
17. 32
18. 16
19. 32
20. 40
21. 24
22. 72
23. 56
24. 16
25. 80
26. 64
27. 48
28. 16
29. 48
30. 80
31. 32
32. 72
33. 40
34. 56
35. 64
36. 24
37. 40
38. 24
39. 56
40. 32
41. 80
42. 48

Page 44

1. 48
2. 32
3. 64
4. 72
5. 24
6. 80
7. 40
8. 16
9. 56
10. 16
11. 64
12. 56
13. 72
14. 48
15. 32
16. 40
17. 80
18. 24
19. 40
20. 24
21. 64
22. 56
23. 72
24. 80
25. 16
26. 32
27. 48
28. 56
29. 16
30. 40
31. 24
32. 80
33. 72
34. 48
35. 64
36. 32
37. 80
38. 64
39. 40
40. 56
41. 72
42. 48

Page 45

1. 54
2. 81
3. 72
4. 90
5. 36
6. 27
7. 63
8. 18
9. 45
10. 90
11. 45
12. 27
13. 54
14. 18
15. 63
16. 81
17. 36
18. 72
19. 54
20. 27
21. 36
22. 90
23. 45
24. 18
25. 81
26. 72
27. 63
28. 54
29. 45
30. 90
31. 36
32. 18
33. 81
34. 27
35. 63
36. 72
37. 90
38. 18
39. 81
40. 63
41. 54
42. 72

Page 46

1. 72
2. 45
3. 18
4. 63
5. 27
6. 90
7. 36
8. 54
9. 81
10. 90
11. 63
12. 18
13. 54
14. 81
15. 72
16. 36
17. 27
18. 45
19. 36
20. 63
21. 18
22. 45
23. 72
24. 27
25. 54
26. 90
27. 81
28. 45
29. 54
30. 18
31. 90
32. 72
33. 36
34. 27
35. 63
36. 81
37. 18
38. 36
39. 72
40. 90
41. 54
42. 27

Page 47

1. 90
2. 81
3. 54
4. 45
5. 27
6. 18
7. 36
8. 63
9. 72
10. 54
11. 63
12. 72
13. 81
14. 27
15. 18
16. 36
17. 90
18. 45
19. 18
20. 45
21. 27
22. 36
23. 90
24. 54
25. 63
26. 72
27. 81
28. 27
29. 63
30. 90
31. 81
32. 36
33. 72
34. 54
35. 45
36. 18
37. 72
38. 45
39. 36
40. 18
41. 90
42. 63

Page 48

1. 30
2. 90
3. 100
4. 80
5. 40
6. 60
7. 70
8. 50
9. 20
10. 100
11. 70

Answer Key

12. 30
13. 60
14. 80
15. 50
16. 20
17. 90
18. 40
19. 90
20. 60
21. 20
22. 30
23. 70
24. 40
25. 100
26. 50
27. 80
28. 60
29. 100
30. 20
31. 80
32. 30
33. 40
34. 50
35. 70
36. 90
37. 40
38. 70
39. 30
40. 60
41. 80
42. 50

Page 49

1. 60
2. 70
3. 100
4. 40
5. 30
6. 20
7. 90
8. 50
9. 80
10. 90
11. 50
12. 60
13. 30
14. 20
15. 70
16. 40
17. 80
18. 100
19. 90
20. 20
21. 40
22. 80
23. 50
24. 70
25. 100
26. 60
27. 30
28. 80
29. 50
30. 60
31. 70
32. 20
33. 40
34. 100
35. 30
36. 90
37. 80
38. 90
39. 100
40. 30
41. 60
42. 50

Page 50

1. 100
2. 50
3. 80
4. 90
5. 40
6. 60
7. 30
8. 20
9. 70
10. 90
11. 30
12. 60
13. 50
14. 70
15. 80
16. 20
17. 40
18. 100
19. 30
20. 60
21. 40
22. 100
23. 20
24. 90
25. 70
26. 50
27. 80
28. 70
29. 60
30. 80
31. 90
32. 40
33. 20
34. 30
35. 100
36. 50
37. 30
38. 60
39. 40
40. 50
41. 100
42. 20

Page 51

1. 64
2. 50
3. 54
4. 70
5. 63
6. 16
7. 27
8. 28
9. 60
10. 30
11. 56
12. 27
13. 90
14. 60
15. 56
16. 36
17. 28
18. 16
19. 36
20. 90
21. 18
22. 50
23. 56
24. 63
25. 48
26. 12
27. 70
28. 32
29. 90
30. 20
31. 49
32. 72
33. 48
34. 60
35. 45
36. 24
37. 50
38. 12
39. 81
40. 28
41. 60
42. 60

Page 52

1. 18
2. 18
3. 56
4. 70
5. 80
6. 81
7. 50
8. 42
9. 32
10. 18
11. 40
12. 80
13. 42
14. 12
15. 63
16. 90
17. 60
18. 28
19. 64
20. 18
21. 50
22. 24
23. 42
24. 60
25. 63
26. 90
27. 32
28. 63
29. 36
30. 28
31. 20
32. 56
33. 45
34. 18
35. 80
36. 81
37. 20
38. 28
39. 40
40. 60
41. 60
42. 54

Page 53

1. 70
2. 27
3. 30
4. 90
5. 48
6. 49
7. 72
8. 32
9. 20
10. 36
11. 18
12. 30
13. 56
14. 80
15. 54
16. 32
17. 49
18. 45
19. 30
20. 42
21. 80
22. 24
23. 80
24. 45
25. 42
26. 81
27. 14
28. 50
29. 32
30. 12
31. 70
32. 24
33. 60
34. 72
35. 42
36. 72
37. 63
38. 54
39. 56
40. 40
41. 45
42. 70

Page 54

1. 32
2. 48
3. 50
4. 54
5. 70
6. 70
7. 54
8. 16
9. 21
10. 90
11. 40
12. 30
13. 63
14. 36
15. 63
16. 24
17. 72
18. 20
19. 63
20. 40
21. 36
22. 64
23. 63
24. 70

25. 40
26. 18
27. 30
28. 64
29. 63
30. 18
31. 45
32. 20
33. 49
34. 48
35. 24
36. 70
37. 48
38. 90
39. 63
40. 48
41. 27
42. 14

Page 55

1. 40
2. 56
3. 18
4. 24
5. 90
6. 24
7. 60
8. 60
9. 49
10. 81
11. 30
12. 18
13. 21
14. 100
15. 56
16. 40
17. 64
18. 42
19. 48
20. 36
21. 21
22. 56
23. 54
24. 54
25. 100
26. 45
27. 20
28. 72
29. 12
30. 35
31. 80
32. 30
33. 63
34. 28
35. 48
36. 48
37. 45
38. 14
39. 60
40. 70
41. 36
42. 64

Page 56

1. 44
2. 22
3. 55
4. 99
5. 88
6. 33
7. 66
8. 77
9. 110
10. 33
11. 77
12. 88
13. 44
14. 55
15. 66
16. 22
17. 99
18. 110
19. 88
20. 110
21. 55
22. 44
23. 66
24. 22
25. 99
26. 33
27. 77
28. 66
29. 33
30. 77
31. 110
32. 44
33. 55
34. 99
35. 22
36. 88
37. 33
38. 44
39. 110
40. 22
41. 88
42. 55

Page 57

1. 55
2. 88
3. 99
4. 33
5. 22
6. 44
7. 77
8. 66
9. 110
10. 55
11. 33
12. 44
13. 110
14. 88
15. 77
16. 99
17. 66
18. 22
19. 55
20. 33
21. 44
22. 99
23. 22
24. 88
25. 110
26. 66
27. 77
28. 55
29. 110
30. 88
31. 22
32. 99
33. 33
34. 44
35. 77
36. 66
37. 22
38. 110
39. 88
40. 99
41. 55
42. 44

Page 58

1. 44
2. 88
3. 22
4. 55
5. 110
6. 66
7. 99
8. 77
9. 33
10. 55
11. 66
12. 44
13. 99
14. 22
15. 88
16. 77
17. 33
18. 110
19. 66
20. 55
21. 99
22. 77
23. 44
24. 22
25. 88
26. 33
27. 110
28. 22
29. 55
30. 99
31. 77
32. 33
33. 44
34. 88
35. 66
36. 110
37. 77
38. 110
39. 66
40. 99
41. 22
42. 88

Page 59

1. 36
2. 84
3. 120
4. 24
5. 72
6. 108
7. 48
8. 96
9. 60
10. 120
11. 60
12. 72
13. 96
14. 84
15. 36
16. 108
17. 48
18. 24
19. 108
20. 48
21. 96
22. 36
23. 72
24. 60
25. 84
26. 24
27. 120
28. 24
29. 72
30. 84
31. 60
32. 108
33. 120
34. 36
35. 48
36. 96
37. 36
38. 48
39. 108
40. 96
41. 24
42. 60

Page 60

1. 96
2. 72
3. 36
4. 48

5. 120
6. 60
7. 24
8. 84
9. 108
10. 72
11. 36
12. 108
13. 24
14. 120
15. 96
16. 48
17. 84
18. 60
19. 24
20. 36
21. 108
22. 60
23. 72
24. 84
25. 120
26. 96
27. 48
28. 108
29. 24
30. 84
31. 36
32. 48
33. 60
34. 120
35. 72
36. 96
37. 36
38. 72
39. 96
40. 120
41. 84
42. 60

Page 61

1. 48
2. 24
3. 96
4. 84
5. 108
6. 120
7. 60
8. 36
9. 72
10. 108
11. 48
12. 60
13. 96
14. 24
15. 36
16. 84
17. 120
18. 72
19. 108
20. 48
21. 24
22. 120
23. 36
24. 84
25. 60
26. 96
27. 72
28. 48
29. 36
30. 108
31. 60
32. 72
33. 24
34. 120
35. 84
36. 96
37. 72
38. 36
39. 120
40. 96
41. 48
42. 60

Page 62

1. 30
2. 32
3. 40
4. 36
5. 24
6. 14
7. 63
8. 90
9. 6
10. 30
11. 110
12. 27
13. 40
14. 48
15. 10
16. 48
17. 28
18. 14
19. 27
20. 10
21. 110
22. 72
23. 36
24. 42
25. 40
26. 16
27. 45
28. 70
29. 40
30. 60
31. 24
32. 99
33. 12
34. 66
35. 4
36. 28
37. 18
38. 90
39. 18
40. 120
41. 48
42. 56

Page 63

1. 63
2. 90
3. 36
4. 40
5. 80
6. 15
7. 48
8. 21
9. 4
10. 99
11. 32
12. 30
13. 32
14. 40
15. 36
16. 21
17. 14
18. 30
19. 70
20. 72
21. 66
22. 6
23. 18
24. 32
25. 30
26. 36
27. 30
28. 21
29. 120
30. 4
31. 20
32. 70
33. 99
34. 24
35. 50
36. 48
37. 32
38. 99
39. 48
40. 60
41. 6
42. 35

Page 64

1. 20
2. 84
3. 24
4. 15
5. 22
6. 18
7. 36
8. 70
9. 48
10. 100
11. 30
12. 64
13. 63
14. 60
15. 54
16. 14
17. 15
18. 8
19. 24
20. 55
21. 27
22. 40
23. 44
24. 63
25. 30
26. 30
27. 56
28. 15
29. 56
30. 24
31. 12
32. 40
33. 96
34. 9
35. 18
36. 8
37. 36
38. 18
39. 35
40. 48
41. 35
42. 32

Page 65

1. 16
2. 12
3. 27
4. 84
5. 18
6. 80
7. 15
8. 50
9. 56
10. 22
11. 54
12. 30
13. 36
14. 33
15. 70
16. 24
17. 30
18. 20
19. 48
20. 49
21. 48
22. 8
23. 40
24. 70
25. 72
26. 30
27. 15
28. 72
29. 72
30. 36
31. 8
32. 12
33. 55
34. 60
35. 84
36. 15
37. 44
38. 63
39. 24
40. 80
41. 10
42. 40

Page 66

1. 27
2. 12
3. 24
4. 24
5. 72
6. 55
7. 12
8. 35
9. 100
10. 32
11. 49
12. 6
13. 18
14. 20
15. 63
16. 88
17. 20
18. 48

19. 48
20. 30
21. 35
22. 90
23. 25
24. 24
25. 63
26. 88
27. 36
28. 30
29. 56
30. 15
31. 16
32. 18
33. 36
34. 90
35. 48
36. 36
37. 14
38. 14
39. 50
40. 72
41. 18
42. 33

Page 67

1. 52
2. 65
3. 26
4. 130
5. 91
6. 78
7. 117
8. 39
9. 104
10. 130
11. 39
12. 52
13. 26
14. 65
15. 91
16. 117
17. 104
18. 78
19. 26
20. 52
21. 39
22. 130
23. 65
24. 117
25. 104
26. 91
27. 78
28. 52
29. 104
30. 130
31. 65
32. 91
33. 26
34. 78
35. 117
36. 39
37. 104
38. 39
39. 65
40. 26
41. 78
42. 52

Page 68

1. 39
2. 52
3. 78
4. 65
5. 104
6. 130
7. 91
8. 117
9. 26
10. 117
11. 65
12. 26
13. 91
14. 130
15. 39
16. 104
17. 52
18. 78
19. 39
20. 65
21. 52
22. 91
23. 26
24. 78
25. 104
26. 117
27. 130
28. 52
29. 117
30. 65
31. 78
32. 26
33. 130
34. 39
35. 91
36. 104
37. 26
38. 52
39. 104
40. 117
41. 65
42. 78

Page 69

1. 28
2. 70
3. 84
4. 140
5. 42
6. 56
7. 126
8. 98
9. 112
10. 140
11. 28
12. 70
13. 42
14. 84
15. 126
16. 56
17. 112
18. 98
19. 56
20. 84
21. 28
22. 112
23. 126
24. 70
25. 98
26. 140
27. 42
28. 70
29. 98
30. 112
31. 140
32. 56
33. 126
34. 84
35. 42
36. 28
37. 70
38. 84
39. 140
40. 112
41. 126
42. 28

Page 70

1. 42
2. 140
3. 84
4. 56
5. 70
6. 112
7. 98
8. 28
9. 126
10. 42
11. 112
12. 56
13. 126
14. 28
15. 70
16. 98
17. 140
18. 84
19. 70
20. 112
21. 28
22. 56
23. 140
24. 126
25. 42
26. 84
27. 98
28. 112
29. 42
30. 126
31. 28
32. 140
33. 56
34. 98
35. 70
36. 84
37. 42
38. 112
39. 84
40. 56
41. 140
42. 28

Page 71

1. 75
2. 105
3. 90
4. 120
5. 135
6. 60
7. 30
8. 150
9. 45
10. 150
11. 75
12. 105
13. 120
14. 45
15. 90
16. 30
17. 60
18. 135
19. 105
20. 75
21. 150
22. 120
23. 135
24. 60
25. 30
26. 45
27. 90
28. 60
29. 120
30. 135
31. 90
32. 75
33. 30
34. 150
35. 105
36. 45
37. 75
38. 60
39. 150
40. 135

Answer Key

41. 90
42. 45

Page 72

1. 60
2. 135
3. 105
4. 75
5. 90
6. 150
7. 45
8. 30
9. 120
10. 60
11. 45
12. 150
13. 75
14. 120
15. 30
16. 135
17. 105
18. 90
19. 60
20. 90
21. 45
22. 30
23. 120
24. 75
25. 150
26. 135
27. 105
28. 60
29. 45
30. 105
31. 30
32. 120
33. 75
34. 150
35. 135
36. 90
37. 150
38. 75
39. 30
40. 135
41. 45
42. 105

Page 73

1. 150
2. 28
3. 72
4. 33
5. 104
6. 108
7. 75
8. 77
9. 52
10. 98
11. 78
12. 112
13. 75
14. 22
15. 108
16. 42
17. 48
18. 130
19. 105
20. 110
21. 39
22. 48
23. 22
24. 120
25. 84
26. 75
27. 108
28. 42
29. 22
30. 130
31. 98
32. 55
33. 120
34. 72
35. 117
36. 56
37. 33
38. 60
39. 120
40. 130
41. 72
42. 98

Page 74

1. 60
2. 26
3. 140
4. 90
5. 99
6. 39
7. 56
8. 120
9. 77
10. 60
11. 105
12. 56
13. 110
14. 96
15. 26
16. 66
17. 45
18. 126
19. 52
20. 120
21. 99
22. 26
23. 60
24. 98
25. 90
26. 36
27. 120
28. 98
29. 66
30. 117
31. 150
32. 65
33. 56
34. 96
35. 33
36. 30
37. 130
38. 126
39. 84
40. 44
41. 28
42. 55

Page 75

1. 44
2. 96
3. 135
4. 26
5. 140
6. 65
7. 84
8. 90
9. 33
10. 70
11. 150
12. 104
13. 33
14. 126
15. 72
16. 98
17. 26
18. 44
19. 90
20. 96
21. 26
22. 108
23. 77
24. 70
25. 150
26. 36
27. 44
28. 126
29. 26
30. 105
31. 110
32. 104
33. 84
34. 60
35. 36
36. 75
37. 72
38. 117
39. 42
40. 77
41. 56
42. 26

Page 76

1. 96
2. 99
3. 140
4. 90
5. 91
6. 22
7. 48
8. 45
9. 65
10. 140
11. 65
12. 72
13. 30
14. 88
15. 56
16. 45
17. 117
18. 84
19. 44
20. 84
21. 130
22. 70
23. 22
24. 108
25. 105
26. 88
27. 36
28. 98
29. 78
30. 60
31. 65
32. 22
33. 126
34. 36
35. 150
36. 112
37. 84
38. 30
39. 44
40. 130
41. 36
42. 70

Page 77

1. 112
2. 32
3. 80
4. 128
5. 160
6. 144
7. 96
8. 48
9. 64
10. 96
11. 64
12. 80
13. 112
14. 48
15. 32
16. 144
17. 160
18. 128
19. 64
20. 32

21. 96
22. 112
23. 160
24. 48
25. 128
26. 144
27. 80
28. 48
29. 96
30. 128
31. 144
32. 112
33. 160
34. 32
35. 80
36. 64
37. 144
38. 112
39. 32
40. 96
41. 160
42. 64

Page 78

1. 112
2. 80
3. 96
4. 32
5. 48
6. 160
7. 128
8. 64
9. 144
10. 160
11. 144
12. 80
13. 48
14. 96
15. 64
16. 112
17. 32
18. 128
19. 64
20. 32
21. 80
22. 112
23. 128
24. 160
25. 144
26. 96
27. 48
28. 64
29. 160
30. 112
31. 128
32. 80
33. 144
34. 32
35. 96
36. 48
37. 160
38. 48
39. 64
40. 80
41. 128
42. 144

Page 79

1. 136
2. 51
3. 153
4. 119
5. 170
6. 102
7. 68
8. 85
9. 34
10. 119
11. 102
12. 170
13. 136
14. 153
15. 68
16. 85
17. 51
18. 34
19. 119
20. 68
21. 170
22. 34
23. 102
24. 85
25. 136
26. 153
27. 51
28. 34
29. 85
30. 102
31. 153
32. 68
33. 170
34. 136
35. 119
36. 51
37. 153
38. 85
39. 136
40. 51
41. 170
42. 68

Page 80

1. 170
2. 136
3. 153
4. 102
5. 34
6. 68
7. 51
8. 119
9. 85
10. 68
11. 119
12. 85
13. 153
14. 136
15. 34
16. 170
17. 102
18. 51
19. 136
20. 102
21. 68
22. 85
23. 34
24. 153
25. 51
26. 170
27. 119
28. 102
29. 153
30. 51
31. 68
32. 170
33. 34
34. 136
35. 119
36. 85
37. 51
38. 136
39. 170
40. 68
41. 102
42. 34

Page 81

1. 180
2. 90
3. 108
4. 36
5. 126
6. 162
7. 72
8. 54
9. 144
10. 108
11. 36
12. 72
13. 90
14. 54
15. 180
16. 162
17. 144
18. 126
19. 36
20. 90
21. 162
22. 72
23. 180
24. 54
25. 144
26. 126
27. 108
28. 90
29. 54
30. 126
31. 144
32. 36
33. 180
34. 108
35. 162
36. 72
37. 54
38. 162
39. 90
40. 144
41. 72
42. 126

Page 82

1. 126
2. 54
3. 36
4. 180
5. 162
6. 108
7. 90
8. 72
9. 144
10. 90
11. 54
12. 100
13. 72
14. 126
15. 144
16. 36
17. 108
18. 162
19. 108
20. 90
21. 144
22. 36
23. 72
24. 162
25. 180
26. 54
27. 126
28. 162
29. 54
30. 144
31. 36
32. 180
33. 90
34. 126
35. 108
36. 72
37. 162
38. 36
39. 72
40. 108
41. 54
42. 144

Answer Key

Page 83

1. 190
2. 152
3. 57
4. 76
5. 133
6. 38
7. 114
8. 171
9. 95
10. 114
11. 95
12. 76
13. 171
14. 152
15. 38
16. 190
17. 133
18. 57
19. 152
20. 171
21. 76
22. 190
23. 95
24. 57
25. 38
26. 133
27. 114
28. 95
29. 171
30. 38
31. 133
32. 152
33. 57
34. 190
35. 76
36. 114
37. 76
38. 171
39. 190
40. 38
41. 114
42. 133

Page 84

1. 114
2. 38
3. 95
4. 76
5. 57
6. 152
7. 171
8. 133
9. 190
10. 76
11. 190
12. 152
13. 114
14. 38
15. 133
16. 57
17. 171
18. 95
19. 133
20. 190
21. 57
22. 76
23. 152
24. 95
25. 171
26. 38
27. 114
28. 76
29. 152
30. 114
31. 38
32. 95
33. 57
34. 133
35. 190
36. 171
37. 114
38. 152
39. 171
40. 95
41. 133
42. 76

Page 85

1. 100
2. 140
3. 200
4. 40
5. 60
6. 80
7. 180
8. 160
9. 120
10. 200
11. 160
12. 80
13. 180
14. 140
15. 120
16. 40
17. 100
18. 60
19. 180
20. 200
21. 60
22. 120
23. 160
24. 140
25. 40
26. 100
27. 80
28. 200
29. 80
30. 40
31. 60
32. 180
33. 120
34. 160
35. 140
36. 100
37. 160
38. 60
39. 100
40. 120
41. 200
42. 40

Page 86

1. 80
2. 60
3. 200
4. 120
5. 40
6. 140
7. 180
8. 100
9. 160
10. 120
11. 80
12. 140
13. 100
14. 160
15. 60
16. 180
17. 200
18. 40
19. 200
20. 80
21. 180
22. 120
23. 60
24. 100
25. 40
26. 140
27. 160
28. 40
29. 140
30. 160
31. 100
32. 80
33. 120
34. 60
35. 200
36. 180
37. 40
38. 180
39. 60
40. 80
41. 160
42. 120

Page 87

1. 180
2. 128
3. 133
4. 40
5. 102
6. 76
7. 90
8. 48
9. 153
10. 120
11. 152
12. 60
13. 119
14. 162
15. 160
16. 85
17. 72
18. 38
19. 160
20. 120
21. 128
22. 72
23. 133
24. 34
25. 60
26. 144
27. 85
28. 126
29. 160
30. 76
31. 100
32. 34
33. 171
34. 108
35. 48
36. 190
37. 40
38. 90
39. 170
40. 96
41. 126
42. 57

Page 88

1. 57
2. 144
3. 200
4. 80
5. 153
6. 36
7. 76
8. 140
9. 96
10. 153
11. 152
12. 96
13. 200
14. 72
15. 34
16. 80
17. 57
18. 126
19. 85
20. 60
21. 32
22. 171

23. 108
24. 80
25. 170
26. 140
27. 136
28. 144
29. 90
30. 133
31. 68
32. 160
33. 36
34. 152
35. 120
36. 57
37. 153
38. 128
39. 108
40. 80
41. 80
42. 133

Page 89

1. 40
2. 48
3. 102
4. 76
5. 180
6. 100
7. 162
8. 136
9. 112
10. 114
11. 160
12. 126
13. 76
14. 34
15. 60
16. 153
17. 128
18. 90
19. 133
20. 160
21. 85
22. 120
23. 162
24. 57
25. 64
26. 190
27. 40
28. 144
29. 160
30. 153
31. 100
32. 96
33. 119
34. 72
35. 38
36. 60
37. 102
38. 126
39. 160
40. 171
41. 68
42. 48

Page 90

1. 126
2. 136
3. 96
4. 60
5. 190
6. 32
7. 180
8. 85
9. 72
10. 171
11. 170
12. 76
13. 100
14. 32
15. 126
16. 136
17. 60
18. 108
19. 95
20. 32
21. 120
22. 64
23. 144
24. 133
25. 51
26. 160
27. 180
28. 102
29. 72
30. 38
31. 51
32. 160
33. 126
34. 95
35. 160
36. 162
37. 160
38. 95
39. 34
40. 180
41. 102
42. 54

Page 91

1. 16
2. 57
3. 80
4. 15
5. 96
6. 80
7. 49
8. 34
9. 162
10. 100
11. 39
12. 70
13. 16
14. 42
15. 99
16. 120
17. 30
18. 24
19. 63
20. 128
21. 90
22. 100
23. 40
24. 15
25. 36
26. 78
27. 99
28. 140
29. 108
30. 42
31. 105
32. 76
33. 12
34. 12
35. 85
36. 24
37. 10
38. 72
39. 21
40. 64
41. 24
42. 100

Page 92

1. 49
2. 108
3. 32
4. 150
5. 8
6. 90
7. 54
8. 20
9. 39
10. 140
11. 32
12. 30
13. 54
14. 34
15. 133
16. 128
17. 15
18. 60
19. 66
20. 81
21. 130
22. 6
23. 32
24. 12
25. 85
26. 70
27. 88
28. 114
29. 25
30. 180
31. 54
32. 42
33. 48
34. 14
35. 14
36. 120
37. 60
38. 128
39. 22
40. 81
41. 20
42. 42

Page 93

1. 84
2. 32
3. 153
4. 12
5. 160
6. 88
7. 50
8. 27
9. 35
10. 6
11. 60
12. 16
13. 108
14. 40
15. 190
16. 91
17. 27
18. 120
19. 14
20. 55
21. 133
22. 120
23. 162
24. 51
25. 72
26. 24
27. 78
28. 12
29. 21
30. 126
31. 40
32. 20
33. 28
34. 40
35. 20
36. 90
37. 80
38. 48
39. 35
40. 84
41. 28
42. 81

Page 94

1. 33
2. 64

Answer Key

3. 42
4. 85
5. 36
6. 20
7. 26
8. 140
9. 152
10. 54
11. 54
12. 60
13. 24
14. 16
15. 50
16. 105
17. 56
18. 60
19. 27
20. 160
21. 140
22. 21
23. 45
24. 16
25. 20
26. 96
27. 72
28. 171
29. 78
30. 68
31. 63
32. 30
33. 30
34. 120
35. 22
36. 70
37. 24
38. 16
39. 60
40. 95
41. 12
42. 98

Made in the USA
Lexington, KY
12 November 2018